RAND

An Evolutionary Approach to Space Launch Commercialization

Brian G. Chow

Prepared for the
Under Secretary of Defense for Acquisition

National Defense Research Institute

This report describes the research and findings of a study on Department of Defense (DoD) policy on commercial space launch services. This research was conducted for the former Office of the Director of Defense Research and Engineering (now, principal Deputy Under Secretary of Defense for Acquisition), and the overall technical cognizance was provided by the office of Dr. George Schneiter, Deputy Director of Defense Research and Engineering (Strategic and Theater Nuclear Forces) [DUSD(S&TNF)]. In DoD Directive 3230.3 of October 1986, the DUSD(S&TNF) was given "the primary responsibility for providing DoD coordination to the DoT [Department of Transportation] on matters arising from the commercial operations of expendable launch vehicles (ELVs) that affect national security interests of the United States." Moreover, the DUSD(S&TNF) is to receive from heads of DoD components reports on "those national security interests of the United States that may be affected by proposed commercial space launch activities."

This study assists the DUSD(S&TNF), as well as the Department of Defense, in developing guidelines for using and strengthening commercial space launch services.

With minor exceptions, the data cutoff date was July 1992, immediately prior to the presentation of the final briefing to the sponsor and representatives from the space launch community of the Air Force, Navy, Joint Staff, and the Office of the Secretary of Defense.

This research was conducted within the Applied Science and Technology Program of RAND's National Defense Research Institute, a federally funded research and development center sponsored by the Office of the Secretary of Defense and the Joint Staff.

CONTENTS

FIGURES

TABLES

KEY RECOMMENDATIONS

The findings and recommendations of this study fall into two groups: Department of Defense (DoD) space launch procurement and DoD steps to strengthen U.S. launch competitiveness. Our analytic results support the choices that the Air Force and the Navy have made since 1985 in the methods of procuring launch services and in the degree of government oversight stipulated in these launch contracts. We further found that the Air Force's upcoming Medium Launch Vehicle-3 (MLV-3) procurement is DoD's most suitable major program to be procured with commercial practices over the next ten years. We recommend that the MLV-3 Request For Proposal (RFP) include commercial launches as an option and that the Air Force consider this option. To help strengthen launch competitiveness, we recommend that DoD concentrate its new launcher development on the most commercially relevant (MCR) range, which is the capability to lift 10,000 to 50,000 lb of payload into low earth orbits (LEOs).

BACKGROUND AND STUDY OBJECTIVE

Historically, the Air Force and the National Aeronautics and Space Administration (NASA) have controlled space launches in the United States. After the January 1986 Challenger accident revealed the danger of heavy reliance on space shuttles, both national and DoD space policies were changed. The new policies direct DoD to purchase commercially available space goods and services to the fullest extent feasible, provided national security requirements are met. In August

1989, the U.S. commercial launch industry provided launch services for the first time on its own to a satellite owner. Furthermore, the National Space Policy Directive of September 1990 called for government agencies, including DoD, to support commercial launch needs. Now that commercial launch services are available, what should the position of the government be with respect to monitoring and controlling the manufacture and launch of space vehicles?

This study's objective is to assist DoD in developing guidelines that both comply with the space policy directives and foster a healthy commercial launch industry. The latter, in turn, would help DoD achieve assured and affordable access to space over a wide spectrum of military commitments in the new strategic environment. The White House and federal agencies have been planning a major overhaul of the way in which the U.S. government procures space hardware. This report should prove useful to that effort.

To facilitate discussion, we introduce our classification of launch procurement contracts. There are three types:

- Government Launches (GL). Traditional procurement. Purchase launch hardware. Cost-plus or cost-plus-like contract and extensive government control and oversight of launcher manufacturing and launch processing. Government makes final decision on launching.

- Commercial Launches (CL). Department of Transportation (DoT) license required. Purchase launch services. Fixed-price type contract and little government control and monitoring except for launch range safety. Contractor makes final decision on launching.

- Commercial-Like Launches (CLL). DoT license not required. Purchase launch hardware or services. Fixed-price type contract. The level of government control and monitoring lies between those of CL and GL. Government makes final decision on launching.

There are three recurring issues in the study:

- How should DoD decide which procurement type (GL, CL, or CLL) is most appropriate for a particular DoD satellite or launch program?

- How can DoD lower its launch cost regardless of the procurement method used?

- How can DoD help to enhance the competitiveness of the U.S. launch industry?

DoD SPACE LAUNCH PROCUREMENT

Since government launches account for two-thirds of the U.S. expendable launch vehicle (ELV) business, many production and quality-control practices the government requires have been used for commercial customers as well. In other words, government requirements, if they are cumbersome and expensive, would increase the cost of doing business even with the commercial customers and lower the competitiveness of the launch industry.

Comparison of Reliability Records

There is a view in some government agencies that GLs are preferable, because CLs or even CLLs are not as reliable. We found, however, that the launch data do not show, with high statistical confidence, that different procurement types result in different launch reliabilities. The average reliability of CLs for expendable launch vehicles in the Delta/Atlas/Titan classes as of July 7, 1992, is 89 percent. With a 95 percent confidence level, the reliability could be as low as 70 percent or as high as 98 percent. For GLs since 1970, the average reliability is 93 percent, with a range from 90 percent to 96 percent at the same confidence level. Whereas the average reliability of CLs is below that of GLs, the CL range blankets the GL range. As of July 7, 1992, there had been 17 successful launches out of 17 CLLs. The reliability range at a 95 percent confidence level is 84 percent to 100 percent. Again, the range overlaps with both ranges of GLs and CLs. Therefore, the reliabilities of GLs, CLLs, and CLs cannot be considered statistically different with 95 percent confidence.

Encouraging Commercialization by Reducing a Manager's Worries

In this report, launch commercialization refers to the use of commercial procurement by the U.S. government in obtaining launch services. It does not mean that the U.S. government will cease providing support to launch research, development, and infrastructure since every launch industry in the world receives such support from its government. There is justifiable concern that the short CL record contains large uncertainties in CLs' true reliability. Consequently, many DoD satellite and launch managers do not wish to reduce government oversight by using CLs, thus slowing the pace of launch commercialization. When the DoD makes future comparative analyses to select a procurement type, we propose that, in the CL procurement option, a satellite/launcher backup be added at the launch site to reduce delay resulting from a launch failure.[1] If a CL procurement, including the backup cost, is still cheaper than a GL or CLL without a backup, CL should be considered. If a backup were found unnecessary, a CL procurement would be even cheaper. Furthermore, we recommend DoD systematically consider the inclusion of launch insurance on launchers and satellites in the launch contract. Buying insurance can lessen a government manager's worry about monetary loss after a launch failure. At the same time, insurance does not have to reduce a contractor's incentive to ensure launch reliability. If his launches fail, his premium will go up, or he might no longer be able to obtain insurance coverage.

Deletion of Undesirable Contract Features

An examination of existing DoD launch contracts revealed several provisions that increase cost or reduce competitiveness. The use of a Fixed-Price Incentive, Firm Target (FPIF) contract in MLV-1 and -2 has resulted in some unnecessary monitoring costs. The use of two prices, target and ceiling, has forced the government to monitor costs and the contractor to provide cost data and to explain cost variances. Had a contract with a single fixed price been used, cost monitoring would have been unnecessary, because the government would pay the same price regardless of actual cost. Another cost-saving measure would be the making of progress payments based on the passage of time, instead of the portion of work accomplished. The

---

[1]A satellite and an unassembled launcher are stored at launch site facilities. In the event of a launch failure, the Air Force has the option to request the backup launch either before or after the post-failure investigation is complete.

latter requires certification and documentation and in a fixed-price contract would not be necessary. Finally, the typical price certification clause, which is meant to let the government benefit from future lower prices charged to other customers, might preclude a contractor from meeting the competition by lowering the price. If the bids are competitive, the government need not restrict a contractor's pricing flexibility in the future.

A Model to Help Decide Whether to Go Commercial

We developed a model for selecting a launch procurement type. The key determinants are satellite cost, launcher cost, number of satellites in the program, launch reliabilities of various procurement types, insurance coverage, and potential savings in using CLs. Applying the model to existing launch programs, we found that our results are consistent with the Air Force's choice of GL for the Titan IV launch of Defense Support Program Block IV Satellites (DSP-BL IV). The Air Force used a CLL procurement for the MLV-1. In retrospect, the Air Force could have saved money by using a CL, because Delta's CL and CLL records have been just as good (11 out of 11 and 15 out of 15, respectively), and CL would have a lower cost because of less government oversight and contractor compliance. On the other hand, the MLV-1 contract was awarded in 1987, well before the first commercial launch in 1989. At the time of the decision, the lack of a CL record made the choice of a CLL procurement for MLV-1 reasonable. We also found the Air Force's CLL procurement for MLV-2 and the Navy's CL procurement for Ultra-High Frequency (UHF) follow-ons to be understandable.

Factors for Deciding Whether to Go Commercial

DoD can use the following set of questions to evaluate whether a particular DoD launch program should be procured commercially (CL instead of CLL or GL):

- Does our procurement selection model indicate a favorable break-even point for commercial launches?
- Do launch vehicles need to be modified?

- How seriously will launch delay affect the timeliness and quality of mission performance?

- If one decided to use commercial launches, would it be feasible and inexpensive to switch to CLLs or GLs if necessary, and if so, at what point?

Recommending That Commercial Procurement Be Included as an Option in MLV-3 RFP

After gaining experience by applying the model to existing launch contracts, we applied it and the above questions to the important, upcoming MLV-3 contract for launching 20 Global Positioning System (GPS) follow-on satellites. The RFP was issued on September 16, 1992, and the contract could be awarded by spring or summer of 1993. The Air Force has incorporated many commercial features into MLV-1 and -2 and is planning to procure MLV-3 in a similar way, namely, CLL. The Air Force could, however, help commercialization even more if MLV-3 were to be procured commercially (CL). The commercial launch industry will benefit from the economies of scale and a longer reliability record associated with 20 additional commercial launches over five years.

Except for small launchers, we found that MLV-3 is the most suitable program for CL and a critical test of launch commercialization. If commercial launch cannot pass this test, it is unlikely to pass any other tests. Also, if the Air Force decided not to pursue MLV-3 commercially, the pace of space launch commercialization would be dramatically slowed.

The Air Force would incur only a limited risk by procuring MLV-3 commercially. We arrived at this conclusion by evaluating the costs and risks according to the four questions above. The procurement selection model gives a favorable break-even point for CLs. The type of launcher suitable for MLV-3 has a CL record that could be as good as CLLs and GLs. The launcher needs only a modest upgrade in capacity, and the upgrade should not alter the launcher so much that the government need worry about the launcher's reliability. Even a few launch failures should not much affect GPS mission performance. Before MLV-3 makes the first launch, the full constellation of GPS (21 satellites plus 3 spares) will have been established. As with other U.S. satellites, they are likely to last beyond their expected lives. If launch failures delay the delivery of GPS follow-ons, the current GPS satellites can serve as substitutes in the interim. Even if the combined GPS and GPS follow-on constellation is incomplete, the system performance would degrade gracefully. An incomplete constellation can perform two-dimensional positional fixing, and three-dimensional would be available for many hours each day.

Our regret analysis showed that if the Air Force purchases insurance only on the launchers, as currently planned, the monetary losses resulting from worse-than-expected CL reliability before switching back to CLLs would be about $100 to $180 million or 9 to 17 percent of the MLV-3 launch program cost. This is a sizable sum, but tolerable. This cost should be considered as the cost of attempting a large-scale launch commercialization by the U.S. government as a whole, not just the Air Force. The government should reimburse the Air Force for a portion, if not all, of the loss. On the other hand, if the CL reliability turns out to be as good as the CLL reliability, the Air Force can save $130 to $250 million or 12 to 24 percent of the launch cost. Whether this savings can be realized is, however, uncertain; it can be determined by requesting such information from the bidders of the MLV-3. The savings are probably not as important as a major successful attempt for commercialization, which, (i) the DoD and the National Space Policy Directive have repeatedly urged, (ii) could make our launch industry more competitive, and (iii) could lead to more launch savings to DoD in the long run.

On the other hand, if the Air Force were also to purchase launch insurance on satellites for MLV-3, the $180 million loss would be recovered. The Air Force might have to pay an extra insurance fee of $50 million, which is over and above the expected cost of launch failures. The net savings to the Air Force could still be from $80 to $200 million if CLs turn out to be as reliable as CLLs. On the other hand, the delay due to a switch-back could still be as long as two years. Short of recommending CLs for MLV-3 outright, we suggest that the MLV-3 RFP contain CL and CLL options as well as different contractor liability options, because the Air Force will need the pricing data to ascertain whether CL can have cost savings and whether the financial impact of launch failures on the government can be limited.

The findings in this report have been communicated to the Air Force. While it finds no objection to some of our recommendations, the Air

Force has decided not to adopt our recommendation of including commercial launch as an option in the RFP. At this stage, unless the new Administration believes our suggestion has merit and the RFP should be modified, it is unlikely that MLV-3 will be procured commercially.

Pace of Launch Commercialization

To us, the key question is not what is usually formulated in the debate: full commercialization now versus no more commercialization ever. In view of the commercial launch record thus far and of the valid concerns many planners have about national security, we consider an evolutionary approach to space launch commercialization to be both feasible and desirable. Moreover, regardless of the pace, there should always be room for a few payload launches to be procured differently. There is already a consensus within the space community that small launchers can be procured commercially. We now recommend that commercial procurement be seriously considered for MLV-3. On the other hand, Titan IVs are not yet ready for commercial procurement. Whether they should be commercially procured in the future depends on the commercial reliability record of Titan IIIs and on how well the commercialization of medium-lift launchers, such as Deltas and Atlases, fares.

DoD STEPS TO STRENGTHEN U.S. LAUNCH COMPETITIVENESS

A Justification for Subsidies

The statement that "the United States will pursue its commercial space objectives without the use of direct Federal subsidies," both in the November 1989 National Space Policy and the February 1991 Commercial Space Policy Guidelines, should be deemphasized in future directives. The statement has been used by the Congress and the Administration in various launch project debates. The distinction between direct and indirect subsidies is artificial. The debates on direct or indirect subsidies can divert the attention from the key issue—namely, whether the proposed subsidies are beneficial to the United States. A justification for subsidies can be based on the simple fact that launch industries in all countries have long been subsidized by their governments. The United States should not be the sole exception. The United States should, however, be willing to work with other countries to reduce and eventually eliminate launch subsidies to attain the long-term goal of "a free and fair market in which U.S. industry can compete," as stipulated in the National Space Policy Directive of September 5, 1990, on commercial space launch policy.

Changes in DoD Launch Demand

The most significant change in DoD launch demand in the past several years has been in the very-heavy-lift launch vehicle (VHLLV) class (above 50,000 lb of payload to low earth orbits [LEO]). Without the heavy Strategic Defense Initiative (SDI) platforms anticipated in the past, we found that most DoD payloads projected for the next twenty or thirty years can be delivered by the existing Titan IVs with upgraded solid-rocket motors, such as the solid-rocket motor upgrade (SRMU) currently under development. DoD's demand for VHLLVs could be very low, whereas NASA's would be much higher. This key difference requires significant compromise in the optimization of engine and other designs for a joint Air Force/NASA launch program.

Launch Developmental Program Must Cover Medium-Lift Vehicles (10,000–30,000 lb to LEOs)

We consider that the most commercially relevant (MCR) range is the capability to lift 10,000 to 50,000 lb of payload into LEOs or 2000 to 10,000 lb into geosynchronous orbits (GSOs). Since commercial competitiveness is a key goal of U.S. technology development efforts, the program should help to improve existing vehicles and to develop a family of new vehicles in the MCR range. Our greatest concern is that any new National Launch System (NLS) family of vehicles might have a lower capacity bound at 30,000 lb for LEOs or 6000 lb to GSOs and thus miss a significant lower portion of the MCR range. Launchers in the 10,000- to 30,000-lb range deliver many important commercial communications, as well as military and civil, satellites (2000 to 6000 lb) to GSOs. Without government financial and other supports to improve and eventually replace vehicles in the MCR range, the U.S. commercial launch industry will suffer greatly and possibly disappear, as competition is intensified by foreign, low-price launch providers and newer launchers.

Government Support Needed to Compete in Commercial Heavy-Lift Vehicles (30,000–50,000 lb to LEOs)

Although Titan III can launch two geosynchronous communications satellites at a time, the difficulties of matching customers' payloads and launch schedule have led Martin Marietta to adopt a policy dedicating each Titan III launch to a single customer. It no longer matches payloads to the same launcher for two different customers. This policy, in essence, positions Titan III launchers for a much smaller market segment of unusually large payloads. Ariane 5, beginning service by the mid-1990s, is expected to include dual-payload launches and to have a launch cost per pound that is 45 percent lower than the already highly competitive Ariane 4. With government support for cost reduction and performance improvement in existing vehicles, the U.S. launch industry believes it can be competitive with Ariane 5 until 2005, perhaps to 2010. We believe that the long-term solution is the early development of new vehicles in this 30,000- to 50,000-lb lift class.

Foreign Partners for Joint Development in Very-Heavy-Lift Launch Vehicles (Above 50,000 lb to LEOs)

This lift class is particularly suitable for international joint development. It will mainly serve scientific space exploration, as opposed to commercial, or even military, purposes. A few future U.S. military platforms might have to be launched by VHLLVs, but a joint development does not preclude the United States from doing so. On the other hand, cost sharing in VHLLV development might leave adequate funds for the development of a new family of launch vehicles in the important MCR lift range (10,000 to 50,000 lb). The United States should actively seek foreign participation in a VHLLV venture.

Matching Funds for Improvements in Existing Vehicles

The NLS program or any successor, as well as other Air Force and NASA programs, is likely to emphasize improvements useful to many launcher types. There will, however, be improvements unique to a specific launcher type. We recommend that matching funds be made available to individual launch providers for improvements of their own launchers and facilities. The matching ratio remains to be specified. We also agree with assessments that U.S. launch facilities are in dire need of repair and upgrade. Otherwise, the obsolete equipment will eventually degrade launch reliability, which is a key determinant in a customer's decision in selecting a launch provider.

ACKNOWLEDGMENTS

The author benefited from Gregory Jones' comments during the first phase of the study and Dana Johnson's participation in the literature search and interviews. The interviews were conducted on a not-for-attribution basis, and they helped the author better understand the views of the participants in the space launch community. The author also found comments from Max Nelson, Katherine Poehlmann, John Hiland, Eugene Gritton, and Richard Hundley useful.

Finally, he appreciated the efforts of Dee Lemke and Janet DeLand, who processed the report with accuracy and speed.

ACRONYMS

AFSLV	Air Force small launch vehicle
ALDP	Advanced Launch Development Program
CDRL	Contract Data Requirements List
CIS	Commonwealth of Independent States
CL	Commercial Launches
CLL	Commercial-Like Launches
COMSTAC	Commercial Space Transportation Advisory Committee
CONUS	Continental United States
CRRES	Combined Release and Radiation Effects Satellite
C-SCSC	Cost/schedule control systems criteria
DMSP	Defense Meteorological Satellite Program
DoC	Department of Commerce
DoT	Department of Transportation
DSCS III	Defense Satellite Communication Systems III
DSP-BL IV	Defense Support Program, Block IV
EIFR	Extra insurance premium rate
ELV	Expendable launch vehicle
FAR	Federal Acquisition Regulations
FIC	Federal investigation cost
FLTSATCOM	Fleet satellite communications
FPIF	Fixed-Price Incentive, Firm Target
GL	Government Launches
GPS	Global Positioning System
GSO	Geosynchronous orbits
GTO	Geotransfer orbit
HLLV	Heavy-lift launch vehicle
IUS	Inertial upper stage

LACE	Low-Power Atmospheric Compensation Experiment
LEOs	Low earth orbits
MCR	Most commercially relevant
MLLV	Medium-lift launch vehicle
MLV	Medium launch vehicle
MSI	Mission success incentive
NASDA	National Space Development Agency
NLS	National Launch System
NUS	No upper stage
P3I	Preplanned product improvements
RFP	Request for proposal
RDT&E	Research, development, test, and evaluation
RME	Relay Mirror Experiment
SDI	Strategic Defense Initiative
SDIO	Strategic Defense Initiative Organization
SEI	Space Exploration Initiative
SRMU	Solid-rocket motor upgrade
STME	Space transportation main engine
STNF	Strategic and theater nuclear forces
TOS	Transfer orbit stage
UHF	Ultra high frequency
VHLLV	Very-heavy-lift launch vehicle
WBS	Work breakdown structure

Chapter One

INTRODUCTION

Traditionally, the Air Force and the National Aeronautics and Space Administration (NASA) have had full control of space launches in the United States. After the January 1986 Challenger accident revealed the danger of heavy reliance on space shuttles, both national and Department of Defense (DoD) space policies were changed. The new policies direct DoD to purchase commercially available space goods and services to the fullest extent feasible, provided national security requirements are met. In August 1989, the U.S. commercial launch industry provided launch services for the first time on its own to a satellite owner.[1] Now that commercial launch services are available, what should the position of the government be with respect to monitoring and controlling the manufacture and launch of space vehicles? Furthermore, the 1991 Commercial Space Launch Policy called for government agencies, including DoD, to support commercial launch needs.[2] On the other hand, both the Directive of November 1989 and the Commercial Space Policy Guidelines of February 1991 instructed that "the United States will pursue its commercial space objectives without the use of direct Federal sub-

---

[1]The first U.S. commercial launch occurred on August 27, 1989, when McDonnell Douglas used its Delta launcher to deliver Marcopolo 1, a British broadcasting satellite, successfully to orbit.

[2]The Commercial Space Launch Policy encourages "technical improvements by directing U.S. government agencies to actively consider commercial needs and factor them into decisions aimed at reducing the costs and increasing the responsiveness and reliability of American launch vehicles." National Space Council, 1990 Report to the President, p. 11.

sidies." How should DoD deal with the issue of direct subsidies and how should DoD support the commercial launch industry?

This study's objective is to assist DoD in developing guidelines that both comply with the space policy directives and foster a healthy commercial launch industry. The latter, in turn, would help DoD achieve assured, responsive, and affordable access to space over a wide spectrum of military commitments in the new strategic environment. The White House and federal agencies have been planning a major overhaul of the way the U.S. government procures space hardware.[3] This report should also prove useful to that effort.

CLASSIFICATION OF LAUNCH PROCUREMENT CONTRACTS

To facilitate discussion, we introduce immediately our classification of launch procurement contracts. As shown in Table 1.1, there are three types:

- Government Launches (GL). Traditional procurement. Purchase launch hardware. Cost-plus or cost-plus-like contract and extensive government control and oversight of launcher manufacturing and launch processing. Government makes final decision on launching.

- Commercial Launches (CL). Department of Transportation (DoT) license required. Purchase launch services. Fixed-price type contract and little government control and monitoring except for launch range safety. Contractor makes final decision on launching.

- Commercial-Like Launches (CLL). DoT license not required. Purchase launch hardware or services. Fixed-price type contract. The level of government control and monitoring lies between those of CL and GL. Government makes final decision on launching.

---

[3]Andrew Lawler, "New Year's Resolution: Overhaul Procurement," *Space News*, November 25–December 1, 1991, p. 3. The interagency team is led by the White House National Space Council with the participation of NASA, the Office of Management and Budget, and the departments of Commerce, Energy, and Defense.

Table 1.1

Classification of Launch Contracts

Contract Type	Cost/Price	Customer Oversight	Launch Decision by	Liability[a]
Government (GL)	"Cost-plus"	Extensive	Government	A
Commercial (CL)	Fixed-price	No	Contractor	A, B, C
Commercial-like (CLL)	Fixed-price	In between	Government	B

[a]In the case of a launch failure, the contractor is liable for
   A: Neither launcher nor satellite
   B: Launcher only
   C: Both launcher and satellite.

In typical GL contracts, the government as a customer does not hold the contractor liable for the loss of launchers or satellites in the event of launch failure. The CLs are structured such that the launch contractor is responsible for launchers and/or satellites or neither. The contractor's liability mandated by the customer depends on the prevailing insurance rate and the customer's risk-averseness. Finally, the government generally holds the contractor liable for launchers, but not satellites, in CLL contracts. When a launch contractor is liable for the loss of launchers or satellites, it can purchase insurance from a third-party insurer or it can self-insure. In the former case, it is likely to include the insurance premium in its bid price, and it is equivalent to the customer purchasing insurance directly from a third party. Further, the premium will have two components. The first component is determined by the size and likelihood of the expected launch loss. The second component covers the cost and profit for the insurer to do business, which we call the extra insurance fee rate, EIFR. If a launch contractor plans to self-insure, it is still likely to charge some EIFR for taking the risk that the actual launch record is worse than expected. Again, the insurance premium will be reflected in the bid price. In this report, we use interchangeably the contractor's liability on launchers and/or satellites and the customer's purchase of insurance on launchers and/or satellites.

Launch commercialization in this report refers to the use of commercial procurement by the U.S. government in obtaining launch services. It does not mean that the U.S. government will cease providing support to the launch research/development infrastructure—

every foreign launch industry receives such support from its government.

There are three recurring issues in the study:

- How should DoD decide which procurement type (GL, CL, or CLL) is most appropriate for a particular DoD satellite or launch program?

- How can DoD lower its launch cost regardless of the procurement method used?

- How can DoD help to enhance the competitiveness of the U.S. launch industry?

STRUCTURE OF REPORT

In Chapter Two, we review government policies that have significant impact on the health and competitiveness of the U.S. commercial launch industry. The critical dependence of the launch industry on the government and the challenges faced by the industry is also discussed.

Chapters Three through Five deal with launch procurement issues. In Chapter Three, we compare the views of various government agencies and private contractors toward launch commercialization. We also identify and compare the commercial features in current DoD launch contracts and discuss why these contracts are structured the way they are.

In Chapter Four, we develop a model that can choose among the three types of launch procurement contracts (GL, CL, and CLL) for particular satellite programs. We first apply the model to existing contracts to gain insight and verification. Then, in Chapter Five, we apply it to the MLV-3, a major upcoming Air Force launch contract, in an attempt to shed light on how the MLV-3 might be procured. We also introduce guidelines for DoD to select commercial launch procurement instead of the more traditional GL and CLL procurement.

In Chapter Six, we outline steps that DoD can take to help strengthen the commercial launch industry and, at the same time, to lower

DoD's cost in the long run. These steps should be beneficial regardless of the type of launch procurement contracts that DoD decides to use in the future. In Chapter Seven,we summarize the findings and recommendations.

Chapter Two

COMMERCIAL LAUNCH HISTORY AND CHALLENGES

COMMERCIAL LAUNCH HISTORY

The Delta launch by McDonnell Douglas in August 1989 was the first U.S. commercial launch. Before that, space launches were considered government launches. NASA and the Air Force traditionally bought launchers from the launch industry and managed the launchings for themselves as well as for all other customers. These customers included other government agencies, foreign governments, and domestic and foreign commercial customers. The U.S. launch industry, therefore, relied critically on the government for business and R&D funding. In 1982, the European consortium, Arianespace, began commercial launches and broke the U.S. space launch monopoly in the Western world. In 1983, the United States decided to launch all government payloads on the space shuttle, and in 1984, passed the Commercial Space Launch Act, which aimed to commercialize expendable launch vehicles (ELVs).[1] Unfortunately, the shuttle launch price was heavily subsidized and set too low for ELVs to compete. Moreover, since all government payloads were to be launched by shuttles, the demand for ELVs would be seriously reduced. These two factors led some launch contractors to close down

---

[1]The Commercial Space Launch Act, its subsequent amendments, and other related documents have been compiled by the Space Commercialization Office, Space Systems Division; see *Commercial Space Launch Act (CSLA) Implementation Handout,* February 15, 1990. A discussion of U.S. space policies prior to June 1988 can be found in Patrick J. Garrity, *United States Space Policy: Review and Assessment,* Center for National Security Studies, LA-11181, Los Alamos National Laboratory, New Mexico, June 1988.

their ELV production lines and to lay off their workers. Fortunately, the Air Force succeeded in ordering ten Titan IVs as backups to shuttles. The production lines of Deltas and Atlases, however, remained in jeopardy.

The Challenger accident in January 1986 and the subsequent Titan, Delta, and Atlas launch failures changed U.S. launch policy drastically. In August 1986, the Reagan Administration directed that no commercial payloads, either domestic or foreign, could be carried by shuttles except for national security and foreign policy reasons and that NASA is prohibited from providing ELV services. Suddenly, the private ELV providers no longer had to worry about the low-priced competition from shuttles or, in fact, NASA competition at all. This probably guaranteed the survival of the U.S. ELV industry, although it does not indicate that the ELV industry will eventually follow commercial, as opposed to government, launch practices. Since government launches account for two-thirds of the U.S. ELV business, many production and documentation practices the government requires are used for commercial customers.[2] In other words, government requirements, if they are cumbersome and expensive, would increase the cost of doing business even with the commercial customers and, thus, lower the competitiveness of the launch industry.[3]

National and DoD space policy directives since 1986 have emphasized that government agencies, including DoD, should purchase commercially available space services to the fullest extent feasible, provided that national security requirements are met.[4]

---

[2]Table 3.3 shows that the U.S. government accounts for about two-thirds of all U.S. launches. This is also the ratio projected by Alan Kehlet, deputy general manager for McDonnell Douglas' Delta Launch Vehicle Division. Daniel J. Marcus, "Trouble Forecast for Launch Business," *Space News*, November 4–10, 1991, pp. 3, 21.

[3]The issue of competitiveness was emphasized in the Commercial Space Launch Act Amendments of 1988. The Act states that "the United States commercial space launch industry must be competitive in the international marketplace."

[4]The DoD Directive of October 14, 1986, covers DoD support for commercial space launch activities. It states that "it is DoD policy to: 1. Encourage the U.S. private sector development of commercial launch operations. 2. Endorse fully and facilitate the commercialization of U.S. Expendable Launch Vehicles (ELVs), consistent with U.S. economic, foreign policy and national security interests...". The Congress made a similar statement in the Commercial Space Launch Act Amendments of 1988: "(1) a United States commercial space launch industry is an essential component of national

That the Air Force increased its Titan IV order from 10 to 23 launches ensured the viability of Martin Marietta's Titan program. Also, the Air Force's signing of the medium launch vehicle MLV-1 launch contract in 1987 and of the MLV-2 in 1988 allowed the Delta and Atlas production lines to remain open. Without the change of U.S. policy in 1986 and the Air Force orders, both McDonnell Douglas and General Dynamics would almost certainly have closed out Delta and Atlas and left the launch business completely. Once the production facilities were dismantled, equipment sold or discarded, and personnel transferred or laid off, resurrecting the launch production capability would be very costly and time consuming, if at all practical. Sole or heavy reliance on shuttles should be considered evidence of a flawed space launch policy.

To date, the launch industry has invested over $600 million in launch modernization and startup.[5] General Dynamics alone has invested $400 million on Atlas' production and launching facilities. Much of this investment remains to be recovered. In February 1991, General Dynamics wrote off $300 million of its $400 million launch investment.[6] A U.S. launch policy containing key provisions that are not supportive of commercial launches would be sufficient to induce one or two major launch providers to leave the commercial launch business. However, the U.S. launch industry can survive on government business alone, although the launch cost to the government could be higher than if the launch business were competitive. A competitive U.S. launch industry that can provide launch services to other countries at a low price also can help to deter other countries from

---

efforts to assure access to space for Government and commercial users; (2) the Federal Government should encourage, facilitate, and promote the use of the United States commercial space launch industry in order to continue United States aerospace preeminence...". The National Space Policy Directive of November 2, 1989, approved by the President, states that "Governmental Space Sectors shall purchase commercially available space goods and services to the fullest extent feasible and shall not conduct activities with potential commercial applications that preclude or deter Commercial Sector space activities except for national security or public safety reasons."

[5]COMSTAC Innovation & Technology Working Group, *FY 1990 Final Report*, a report to the Commercial Space Transportation Advisory Committee (COMSTAC), October 18, 1990, pp. 3, 6.

[6]Debra Polsky, "Mission Boosts Atlas Revenues, But Second Write-off Possible," *Space News*, December 16–22, 1991, p. 18.

developing their own space launch vehicles or ballistic missiles; thus, it helps to slow missile proliferation.

The Worldwide Launch Market

Since the Reagan Administration's decision in August 1986, commercial payloads, with few exceptions, are not carried by shuttles. DoD also decided that its payloads will be carried by expendable launch vehicles (ELVs) except when a payload can be delivered only by a shuttle. Thus, ELVs have become the primary vehicles for delivering commercial and military payloads worldwide.

Table 2.1 shows the domestic and foreign ELVs that are currently or soon to be available. Two points can be made. First, the United States has long maintained the policy that U.S. government payloads can be launched only by U.S. launch providers. Since U.S. launch providers tend to serve different lift classes, the domestic competition among them has not been fierce.

In small launchers competition, Orbital Sciences Corporation serves the small launcher market with very limited competition from LTV or Martin Marietta. The Scouts and Titan II have a launch cost about double those of Pegasus and Taurus, respectively.[7] Other small launchers, such as AMROC's ILV and SSI's Conestoga, have not garnered enough business to ensure their viability. Their developers are new, struggling firms that cannot withstand any prolonged drought in financial support. Orbital Sciences' main concern is the possibility of converting surplus strategic missiles into space launch vehicles by its potential competitors, such as Lockheed. In the longer term, foreign entries into the small launcher market could also be serious because many are subsidized by their governments. Delta launchers serve in the 9000–11,000 lb (to low earth orbits [LEOs]) class, Atlas in the 13,000–19,000 lb class, and Titan III & IV in the 30,000–50,000 lb class. The upcoming MLV-3 launch contract will, however, place

---

[7]Strictly speaking, we should refer to these as launch prices rather than costs. The term "launch cost" is, however, more commonly used in the literature. Also, launch prices charged by providers are launch costs to customers. Thus, we will use launch cost and launch price interchangeably in this report.

Table 2.1

Expendable Launch Vehicles in the World

Nation/ Consortium	Company	Vehicle Family	Version	Configuration	Vehicle Availability	Available Commercially?	Lift in lb (Orbit indication = 28 degrees)			Launch Price (1990$M)	Source
							100 n mi LEO	GTO	GSO		
U.S.	LTV	Scout	I		Since 1960s	Yes	570			10–13	1,3
			II		Early 1990s	Yes	1100			16	3
	Orbital Sciences	Pegasus			1989		1100			7	3
		Taurus			Early 1990s		3700	830		16	3,7
	McDonnell Douglas	Delta	II	6925	February 1989	Yes	8780	3190	1600	42–	1
				7925	1990	Yes	11110	4010	2000	50	1
	General Dynamics	Atlas	I		July 1990	Yes	13000	5150		65–70	1
			II		1992	No	14950	6100		70–80	1,7
			IIA		1992	Yes	15700	6400		80–90	1
			IIAS		1992	Yes	19000	8000		110–120	1
	Martin Marietta	Titan	II		September 1988	No	4200			35–40	1
			III		1989	Yes	30500	11000		145–	1
			III	With SRMU	Early 1990s	Yes	38000			155	1,7
			III	With TOS	1992	Yes		13000		190–200	1
			III	With IUS	1989	Yes			4200	245–	1
			III	With IUS & SRMU	Early 1990s	Yes			5000	255	1,7
			IV	With NUS	1989	No	39000			168–180	1,8
			IV	With NUS & SRMU	Early 1990s	No	49000	15000	6600	177–240	1,7,8,9
			IV	With IUS	1989	No			5200	168–180	1,9

Table 2.1—continued

Nation/Consortium	Company	Vehicle Family	Version	Configuration	Vehicle Availability	Available Commercially?	Lift in lb (Orbit indication = 28 degrees) 100 n mi LEO	GTO	GSO	Launch Price (1990$M)	Source
			IV	With Centaur	1990	No			10200	217–260	1,8
			IV	With Centaur & SMRU	Early 1990s	No			13500	225–320	1,7,8
European Space Agency	Ariane-space	Ariane	1		Last launch 1986	No	10670	4000			2
			2		Through 1989	Yes	11000	4900			2
			3		Through 1989	Yes	12760	5700		100	2,4
			4		1989	Yes	17600	9250	4850	120	2,4
			5		1995	Yes	42000	15000	8820	107–195	2,6
Japan	NASDA	N	I		Last launch 1982	No	2645		287		2,4
			II		Last launch 1986	No	4409		772	70	2,4
		H	I		Through early 1990s	No	6614	2300	1213	120	2,4,7
			II		Early 1990s	Yes	17640	8800	4850		2,4,7
PRC	Great Wall Industry	Long March	1		Through 1990s	No	800	1300			2
			2		About 1990	Yes	5500	2756		30	2,5
			3		1990	Yes		5500		30	2,5
			2E		1990	Yes	19800	7040			5
CIS	Glavcosmos	Proton	SL-13		Through 1990s	Yes	44700	10000		35	2,4
			SL-12		Through 1990s	Yes	44090			30	2,4
		Zenit	SL-16		Through 1990s	Yes	33069		1433	30	2,6
		Energia	SL-X-17		1987	Yes	220460	50000		30	2,5

Table 2.1—continued

NOTES:

- In the Launch Price column, xx- on one line and xx on the line below means that the corresponding launch configurations on both lines span the price range. For example, Delta II 6925 and 7925 have a price range of $42 to $50 million.

- LEO = low earth orbit; GTO = geotransfer orbit; GSO = geosynchronous orbit; CIS = Commonwealth of Independent States; NASDA = National Space Development Agency; TOS = transfer orbit stage; IUS = inertial upper stage; and NUS = no upper stage.

- In this report, we define that light-lift launch vehicles (LLLVs) can lift no more than 10,000 lb of payload to LEOs; medium-lift launch vehicles (MLLVs) 10,000 lb to 30,000 lb; heavy-lift launch vehicles (HLLVs) 30,000 lb to 50,000 lb; and very-heavy-lift launch vehicles (VHLLVs) more than 50,000 lb.

- Recent test failures on the Solid Rocket Motor Upgrade (SRMU) have caused delay and possibly cancellation. The availability dates shown in the table are target dates estimated before the failures.

- The U.S. space shuttle has a lift capacity to LEO of 51,000 lb, which can be used for a comparison with ELV numbers.

- The launch costs of Scout I and II, Pegasus, and Taurus were originally given in 1989 $ and have been converted to 1990 $ by using an implicit price deflator of 1.04.

- Glavcosmos has been marketing additional Soviet launchers, such as Soyuz (15400 lb to LEO), Vostok (10400 lb), Molniya (3300 lb), Tsyklon (8800 lb) and Cosmos (990 lb). See Source #5, p. 53.

- Many numbers in the table have not been rounded to their significant figures.

SOURCES:

1. Karen Poniatowski, *U.S. Civilian Government Expendable Launch Vehicle Payload Compendium,* 5th ed., NASA Office of Space Flight, June 1990.
2. Karen Poniatowski, *Expendable Launch Vehicle Capabilities, Constraints, and Costs,* NASA Office of Space Flight, March 9, 1989, pp. 4, 5.
3. Karen Poniatowski, *Compendium of Small Class ELV Capabilities, Costs, and Constraints,* NASA Office of Space Flight, undated (after July 1989), pp. 7, 9.
4. Albert Wheelon, *Space Policy: How Technology, Economics, and Public Policy Intersect,* Massachusetts Institute of Technology, 1989, p. 46.
5. Berner Lanphier and Associates, Inc., "Assessment of Foreign Activities That Affect NASA's Commercial Space Program," briefing, May 1, 1990, pp. 36, 53.
6. COMSTAC Innovation and Technology Working Group, *FY 1990 Final Report,* a report to the Commercial Space Transportation Advisory Committee (COMSTAC), October 18, 1990, pp. 10, 21.
7. Updated by the author.
8. Launch vehicle cost data provided by the Air Force Space Systems Division, June 29, 1992.
9. Assume Titan IV with IUS costs performs the same as Titan IV with NUS.

Delta and Atlas into direct competition, because the required lift capability is near the high end of Delta's lift capability and the low end of Atlas'. Of course, other launch providers could also bid with new or modified launchers, as in the case of MLV-2 competition. Titan IV is currently in an enviable position. It has no domestic competition, at least until the end of the decade. Moreover, U.S. policy disallows foreign launch providers to bid on the launches of U.S. government payloads.[8] In such an environment, Martin Marietta has a much stronger motivation to satisfy the government requirements than to lower the cost. The implication is that any initiatives for launch cost reduction in that class would have to come from the government.

Second, although the competition among domestic launch providers might not be fierce, Arianespace has captured about 60 percent of the Western launch providers' commercial market (all except U.S. government launches), and the situation is worsening as Japan and nonmarket providers, such as the Peoples Republic of China (PRC) and the Commonwealth of Independent States (CIS), enter the market.[9] Fortunately, as reflected in the cost and performance columns of Table 2.1, our allies do not currently have a clear cost advantage over us even though their technology and infrastructure are more modern. Arianespace has indicated that Arian 5 could be priced 45 percent lower than Adrian 4 in dollars per pound.[10] By taking actions quickly (some of which will be recommended in this report), the United States can prevent further erosion of our launch business to other countries charging market launch prices. Our dealing with nonmarket providers includes continued reliance on setting launch

---

[8]The National Space Policy Directive of September 5, 1990, on commercial space launch policy, plans "the continued use of U.S.-manufactured launch vehicles for launching U.S. Government satellites" in the near term (until the year 2000).

[9]It is likely that any market system established in the CIS will eventually cover the pricing of its space business. It is also likely that the Western launch providers will be dealing with individual Soviet republics directly, instead of the CIS, on space matters. Although the political reorganization continues to generate uncertainties about the Soviet space apparatus, the current thinking is that as many as five space agencies will emerge. Three agencies will be in Russia, one in Kazakhstan and one in Ukraine. All three republics are in the new CIS. Vincent Kiernan, "Five Space Agencies Emerge from Soviet Chaos," *Space News*, December 16–22, 1991, p. 3.

[10]COMSTAC Innovation & Technology Working Group, *FY 1990 Final Report*, a report to the Commercial Space Transportation Advisory Committee (COMSTAC), October 18, 1990, p. 10.

limits to place U.S. satellites in orbit until these governments develop a free market or agree to use some sort of market pricing system.

CHALLENGES FACING THE U.S. COMMERCIAL LAUNCH INDUSTRY

Essentially, the United States has a single launch industry serving the needs of government clients and commercial customers. Since the U.S. government is a major launch customer, there is a tendency for the launch industry to focus its attention in meeting the government requirements and to think that, even without commercial business, the government alone can keep the launch business afloat. This might well be true. The government should, however, recognize that a noncompetitive launch industry will eventually make the government pay a high price for launching. The launch industry is facing two challenges and the government's help is crucial for the industry to meet them.

The first challenge is the need to upgrade the U.S. launch infrastructure. To serve the military need, the U.S. space launch program has concentrated from the start on high performance instead of low cost. A performance orientation tends to push for the last increment in lift capability. Consequently, the safety margin is slimmer and the testing procedure more elaborate. Ironically, time-consuming procedures do not even serve the military well. Our launch processing time is typically 60 days or more for the Delta class and above and is especially ill-suited for emergency and wartime launches. On the other hand, foreign competitors, such as the European Space Agency and Japan, have concentrated their launch development on commercial applications and on cost. In spite of their commercial orientation, these foreign competitors do not, at present, have a definite cost advantage over U.S. launch providers.

The picture becomes dimmer when we look into the future. Our major competitor, Arianespace, has been investing heavily on Ariane 5, with a goal of reducing the cost per pound by 45 percent. The joint Air Force/NASA program for the National Launch System (NLS), which can have applications in commercial launchers, is far behind in its development stage and financial support and was recently cancelled.

Worse yet, the U.S. launch infrastructure is of 1950 vintage—40 years old. Some of the equipment is so old that obtaining replacement parts is a major problem. Thus far, through the ingenuity of the processing and maintenance crew, there have been no major mishaps due to old equipment. Unfortunately, this achievement buries the urgent need to replace the obsolete equipment in the very near term.

Obsolete equipment will eventually degrade launch reliability. Since a satellite often costs considerably more than the launcher, and the value of a satellite's service is even higher, a customer's decision in selecting a launch provider hinges heavily on the reliability record. A poor reliability record would cause the United States to lose the commercial launch business. Considering also that a launch failure itself would easily cost over $100 million for the launcher and satellite plus interrupted or delayed satellite service, we believe that preventing a failure might well be cheaper than curing the results of one.

DoD has expressed a desire for more responsive launches to meet new contingency support requirements. Upgrading existing launchers and their infrastructure is a solution for the near future.

The second challenge relates to foreign launch providers, who are gaining experience and investing heavily. A particular concern involves providers, such as PRC and CIS, who still charge nonmarket launch prices. These providers have been undercutting the West by offering as much as 50 percent discount for space launches. For the near future, we can limit their market penetration by specifying the number of launches involving U.S.-technology-based satellites that these countries can launch over a period of time. The United States needs, however, a solution for the long run, because U.S. patents will expire and, in any case, denial of access to low-priced launchers inhibits the ability of U.S. satellite producers to compete in the global satellite market. Foreign satellite makers, being able to use low-priced launch services, can offer a lower-priced delivery-to-orbit package to customers. Moreover, there is a growing global oversupply of launch services.

Since every foreign launch industry is subsidized by its government, the U.S. government will have to continue to do the same if its industry is to remain competitive. To help the launch industry the most, the U.S. government should minimize imposing requirements and thus costs to launcher manufacturing and processing. At the same time, let the industry benefit from the large government launch demand and R&D funding support.

Chapter Three

UTILIZATION OF COMMERCIAL LAUNCH SERVICES

COMPARISON OF RELIABILITY RECORDS OF THREE TYPES OF LAUNCHES

A key factor that determines whether commercial, government, or commercial-like procurement will be used is launch reliability. Reliability could be affected by the type of procurement, because different types call for different levels of government oversight. Does the lack of Air Force supervision lead to lower launch reliability?

Tables 3.1 and 3.2 show U.S. launches as of July 7, 1992. This is our last data update, immediately before our final briefing to the project sponsor. There were 19 commercial launches and 17 commercial-like launches. The commercial launches were made with Delta, Atlas, and Titan III, whereas the commercial-like launches were made with Delta II for the Global Positioning System (GPS) satellites under the Medium Launch Vehicle-1 (MLV-1) program and Atlas II for Defense Satellite Communication System III (DSCS III) under the MLV-2 program.

Since 1989, there also have been 12 government launches. The 48 launches in this period are approximately equally divided among the three launch procurement types (Table 3.3). There were two launch failures, both of the commercial launch type—one with Titan III and one with Atlas I (Table 3.1). It is, however, inappropriate to conclude quickly that commercial launches are less reliable.

Because our objective is to compare the reliability records of commercial, commercial-like, and government launches in the Delta,

Table 3.1

Comparison of U.S. Launch Reliability Records
(January 1, 1989 to July 7, 1992)

Launch	No. of Successes	No. of Launches	Reliability (percent)
CL			
Delta	11	11	100
Atlas	4	5	80
Titan	2	3	67
Total	17	19	89
CLL			
Delta	15	15	100
Atlas	2	2	100
Total	17	17	100
GL			
Delta	2	2	100
Atlas	1	1	100
Titan	9	9	100
Total	12	12	100

Atlas, and Titan classes, we excluded the records of smaller launchers such as Scout and Pegasus. Since 1989, there have been 17 successful U.S. commercial launches and two failures (Table 3.2). The average reliability is 89 percent.[1] Still, 19 launches is a small number, and some initial failures are understandable. With a 95 percent confidence level, we found that the reliability could be as low as 70.4 percent or as high as 98.1 percent. For government launches, we have a much longer historic record. Since 1970, there were 278 government launches with 259 successes.[2] The average reliability for U.S. government launches is 93.2 percent, with a range of 90.2 percent to 96.1 percent. Although the average reliability of commercial launches is below that of U.S. government launches, the range for commercial launches actually blankets that of government launches

---

[1]To facilitate the replication of our calculations by other researchers, we have not rounded many numbers in this report to their significant figures.

[2]The three main families of expendable launchers in the United States—Atlas, Delta, and Titan—were started in the late 1950s or early 1960s. We used launch reliability data after 1970, because we did not want to include failures during developmental stages. Those failures are not representative of the established families.

Table 3.2

Launch Reliabilities Appear to Differ with Launch Type
(as of July 7, 1992)

Launch	Since	No. of Successes	No. of Launches	Reliability (percent)
U.S. CL	1989	17	19	89
U.S. CLL	1989	17	17	100
U.S. GL[a]	1989	12	12	100
U.S. GL[a]	1970	259	278	93
Ariane	1979	45	50	90
Japan	1975	24	24	100

[a]To obtain data on comparable launchers, we excluded Atlas Es. They are generally considered to be dissimilar to Atlas/Centaurs, which serve all three types of launch procurement contracts.

(Figure 3.1). As of July 7, 1992, there had been 17 successful launches out of 17 CLLs. The reliability range at 95 percent confidence level is from 84 percent to 100 percent. Again, the range overlaps with both ranges of GLs and CLs. Therefore, the launch data do not show, with high statistical confidence, that different procurement types result in different reliabilities.

Also shown in Table 3.2 and Figure 3.1 are the records for Ariane launchers and Japanese launchers. The Ariane launches in particular can be considered as commercial launches. Again, their reliability ranges blanket those of U.S. government launches. One cannot say that, at the 95 percent confidence level, foreign launches and U.S. government launches have different reliabilities.

In addition to the average reliabilities of various procurement types, another key determinant for selecting CL, CLL, or GL is how likely it is that the future reliability could deviate from the historic record and by how much. Prospective launch users are worried that the reliability for launching their payloads may turn out much worse than the past record. Many users select government launches because of their long record.

The average reliability of government launches is 93.2 percent, which compares with 89 percent and 100 percent for commercial and commercial-like launches, respectively. The differences range from

Table 3.3

**U.S. Launch Frequencies in Delta/Atlas/Titan Classes
(January 1, 1989 to July 7, 1992)**

Launch	Customer			Total
	DoD	NASA	Commercial	
CL	1.5[a]	.5[a]	17	19
CLL	16	1	0	17
GL[b]	10	2	0	12
Total	27.5	3.5	17	48

[a]The Combined Release and Radiation Effects Satellite (CRRES) launched on July 25, 1990, was a joint Air Force/NASA program and, therefore, counted as half DoD and half NASA.

[b]To obtain data on comparable launchers, we excluded Atlas Es. They are generally considered to be dissimilar to Atlas/Centaurs, which serve all three types of launch procurement contracts.

RAND#460-3.1-0593

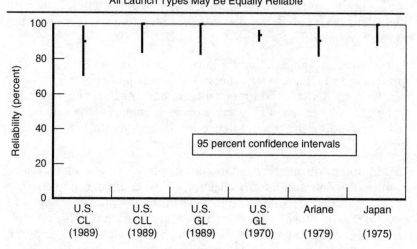

Figure 3.1—Comparison of Launch Reliability Records
(as of July 7, 1991)

4 to 7 percentage points. On the other hand, at a 95 percent confidence level, the CL and CLL reliabilities could be as low as 70.4 percent and 83.8 percent, whereas the GL reliability could only be as low as 90.2 percent. The differences in lower bounds are larger and range from 6 to 20 percentage points. For very risk-averse users, it is understandable that government launches would be a safer choice, because the chance for government launches to degrade greatly is lower.

The reluctance to trade confidence in reliability for cost savings, even if they exist, has slowed the pace of U.S. launch commercialization. Unfortunately, the lack of high confidence is inevitable for any new endeavor. A new venture will always start with a small number of data points and a low level of confidence concerning its outcome. This is the fundamental dilemma facing planners who intend to implement the DoD and national space policy directives of using commercial or commercial-like launches to the maximum extent possible.

There are two ways to alleviate this fundamental concern. The first approach is based on the observation that even less reliable individual launches can result in a launch system of high reliability. We want to determine whether additional measures can be taken on CLs and CLLs to enhance the overall reliability of the satellite launch program in spite of the lack of high-confidence reliability of individual launches. For example, we might add a provision in the CL or CLL contract for the availability of a backup payload and launcher at the launch site. Then, if a CL or CLL failure occurs, the user will have the option to use the backup to avoid long delay. In contrast, a GL contract without a backup provision might end up having a lower confidence level in overall system reliability. For certain DoD satellite programs, we will show that a CL or CLL with backup could be cheaper than a GL without backup.

Second, insurance might alleviate the monetary loss from launch failures. Moreover, an insurance premium that an individual launch provider pays reflects the expected reliability of its launchers. The insurance premium is a way to use market forces to adjust for the fact that different launch providers have had different CL or CLL reliability records. For example, Delta has made 26 CLs or CLLs, whereas Atlas and Titan have had only 7 and 3, respectively. If one combines the CL and CLL records of Delta, 26 successes out of 26 have been attained. At a 95 percent confidence level, the reliability is unlikely to be lower than 89.1 percent. In contrast, the government launch reliability is unlikely to be lower than 90.2 percent. These two numbers are similar. If no insurance on the launcher and payload were required in the launch contract and if reliability is a major figure of merit, Delta would have a high edge over other launchers. The requirement for insurance could help other launch providers, especially those with a short or poor launch record, to win a launch contract, because insurance alleviates the user's concerns about monetary losses.

VIEWS AND CONCERNS REGARDING LAUNCH PROCUREMENT AND COMMERCIALIZATION

To determine the pace and extent of launch commercialization, we first need to understand the views of involved parties. A better understanding helps us recommend a policy that addresses their concerns. In our discussion below about the views of various agencies and contractors, note that planners within the same agency or firm do not necessarily have similar views. Our main purpose is to list the major views and concerns; it is less important who actually holds what view.

In this section, we will also describe recent major DoD launch contracts. The three major Air Force launch contracts are Titan IV, MLV-1, and MLV-2, and the major Navy contract is the Ultra High Frequency (UHF) Follow-on. A much smaller but typical contract, the Strategic Defense Initiative Organization's (SDIO's) LACE/RME (Low-Power Atmospheric Compensation Experiment/Relay Mirror Experiment) launch contract, will also be discussed. The upcoming major Air Force MLV-3 contract will be examined in Chapters Four and Five after we introduce a quantitative method and a set of criteria for selecting which procurement type to use.

Air Force

An address by Martin Faga, former Assistant Secretary of the Air Force for Space, amply reflected the Air Force's position on commercial space launches.[3] He observed that, for a typical DoD space program, launch support accounts for 25 percent of the total program cost. About half of the launch support is related to launch procurement, and the remaining half is equally divided between launcher development and launch operations. Thus, launch procurement accounts for only about 12 percent of a space program's cost, and launch operations, only 6 percent. On the one hand, Faga believes that commercially reasonable practices should continue to be encouraged in every procurement. Examples include eliminating unneeded contract provisions and certifying entire processes instead of inspecting launcher components on the assembly line. On the other hand, he believes that DoD should not rely on commercial practices for launch operations and control, because some DoD missions are so critical that it is not worthwhile to trade a small cost savings for an increase in the risk of launch failure. It has also been reported that the Air Force believes that commercial firms can launch small payloads, but its officials are uncomfortable with any plan that uses commercial services for large, expensive payloads.[4]

It seems that former Assistant Secretary Faga is more comfortable in allowing more commercial practices in the manufacturing of launchers than in the processing of the launchers at the launch sites. Three points can be made. First, since the Air Force owns the launch sites and many of the launch facilities, commercializing the launch processes would affect the Air Force more than commercializing launcher production. Moreover, since launcher manufacturing costs about twice as much as launch operations, the potential cost savings in the former could be higher, at least in the near term.[5]

Second, there seems to be an implicit assumption that commercial launch will incur a higher launch failure risk. This is not borne out by

---

[3]Martin Faga, "Commercial Space: A Pentagon View," *Space News*, August 19–25, 1991. Remarks adapted from an address to the Maryland Space Business Roundtable in Greenbelt, Maryland, on July 23, 1991.

[4]"NASA, DoD buck trend to 'buy commercial'," *Space Business News*, May 28, 1990, p. 7.

[5]In fact, since the Air Force charges only direct costs to the contractor for launch site services, the launch processing cost to the contractor can be less than a quarter of the launcher manufacturing cost. In other words, the contractor's launcher manufacturing cost can be about four times that of the launch processing cost.

our analysis of the reliability records above. Former Assistant Secretary Faga could also be concerned that the commercial launch record is too short to provide high confidence. This is true in general, but there are two exceptions. The commercial Delta is emerging with an impressive (26 out of 26 launch successes) and sizable record. Also, if commercial launches were never given a chance, they could never establish a launch record with any level of high confidence. One strategy is not to make an abrupt and complete conversion from government-controlled launches to contractor-controlled launches. Rather, we could start with certain classes of launches first.

Third, although launch support accounts for only 25 percent of the total space program cost, the absolute dollar value is still large and of great commercial importance. There are other noneconomic concerns if the U.S. launch industry is not competitive in the world market. For example, the U.S. argument that it is necessary to discourage Third World countries from entering the space launch market to prevent proliferation of ballistic missiles could be interpreted as a way to protect the noncompetitive domestic launch industry. At the same time, our high-cost launches would indirectly encourage Third World ballistic missile development, because these countries might believe that they could win some space launch business and lower their net investment for ballistic missile development.

Others at the Air Force have suggested that small launchers, such as Pegasus, are suitable for commercial launches. The payloads for small launchers are generally cheaper and less critical and one could afford to take a higher risk. Other government agencies generally agree with this view. The difference comes when one starts to discuss whether larger launchers, such as Delta, Atlas, or even Titan, which carry more expensive and vital payloads, are now ready for commercialization and when they should be commercialized. Planners at the Air Force generally feel uncomfortable losing control of launchers, with the possible exception of small launchers. This view is reflected in the statement of General Robert Dickman, deputy director of Air Force space acquisition.[6] He said, "since we can't insure our payloads, we must do everything reasonable and prudent

---

[6]"Air Force rejects commercial launches," *Military Space*, July 29, 1991, p. 6.

to ensure they are delivered safely to orbit."[7] He further cited the need to launch three GPS satellites in a hurry after the Iraqi invasion of Kuwait and, therefore, the need to have direct control over the launch schedule. Again, the implicit assumption is that commercial launch is less reliable or certain. As to the need for emergency launches, we agree that any commercialization scheme should allow DoD to make such launches.[8]

Finally, the Air Force generally believes that government oversight provides a cost/reliability benefit. Our comparison of reliabilities of GLs, CLLs, and CLs does not show such benefit exists with high confidence. In any case, if it exists, a CL contractor can still take advantage of it by hiring the Air Force, the Aerospace Corporation or others to perform the oversight.

Described below are the current major Air Force launch contracts.

Titan IV. Immediately after the 1983 U.S. decision that all government payloads would be launched on shuttles, the Air Force became concerned about this sole reliance. By 1985, the Air Force contracted with Martin Marietta for ten launch vehicles, later named Titan IVs. These launchers were to be used to launch some of the highest priority military satellites, such as a ballistic missile attack early warning system, MILSTAR communication satellites, and other classified payloads.[9] After the Challenger accident in 1986, it was apparent that the Air Force's insistence on having ELVs was a wise move. In the same year, the Air Force increased the order to 23 launchers. Later, the number further increased to 55, with 41 vehicles to be launched through the mid-1990s. As late as 1990, the Air Force plan after 1995

---

[7]It is not clear to us what he meant by "can't insure our payloads. He could mean that the launch insurance premium on satellite is too high for him to pay. He could also mean that a launch failure will cost more than financial losses. Launch and mission delay cannot be compensated. This report will address both issues—what is the typical launch insurance premium? Will launch delay affect a satellite constellation's performance significantly?

[8]The impact of interjecting government emergency launches on commercial launch services depends on the number of launch pads and the length of on-pad time for each launch, because the launch pads are typically the launch processing bottleneck. Since the U.S. launch industry is not operating at maximum capacity, it can accommodate some emergency launches.

[9]U.S. General Accounting Office, *Space Launch: Cost Increases and Schedule Delays in the Air Force's Titan IV Program*, May 1990.

was to procure and launch about ten Titan IV launchers a year. The drastic changes in the geopolitical situation and the general reductions in military procurement budgets have led to a lower rate of eight or even five launches a year.[10] The Titan IV program will remain important for payloads above the Delta and Atlas class. A key difference between the Titan IV and the other launch programs (to be discussed below) is that the Titan IV launchers are being used for a wide variety of payloads, as opposed to one main type of payload, such as GPS on MLV-1 Delta launchers and DSCS on MLV-2 Atlas launchers. Also, Titan IV payloads are generally more expensive. The constellation established by Titan IV launches often consists of fewer satellites than those established by Deltas and Atlases. We will explain in Chapters Four and Five why these attributes are important in deciding whether a particular launch program should be procured differently in the future.

Another important element in a procurement contract is insurance. Is the satellite owner expected to be self-insured? Or is the launch provider expected to obtain insurance on the launcher and/or satellite? Let us examine the insurance or free reflight provision in the Titan IV program. In the Titan IV contract, each launch failure would take $45.3 million from Martin Marietta's combined profit and incentive pool of $675 million.[11] Since each Titan IV launch costs $221 million (see Table 4.4), the penalty amounts to only 20 percent of the launch cost, and the Air Force would still have to assume the remaining 80 percent. The penalty induces the launch provider to pay added attention to high launch reliability. The Air Force, however, is justifiably concerned about its 80 percent share of the reflight cost and the full share of the satellite replacement cost of, say, $200 million (see Table 4.4). Moreover, the delay in satellite service is another major concern. From the perspective of a satellite program manager, the Titan IV contract is well approximated by our no-insurance case (to be discussed later), because the equivalent amount of insurance coverage for Titan IV is small.

---

[10]Edward Kolcum, "Reduced Military Budgets Revamp Titan 4 Production and Launch Program," *Aviation Week & Space Technology*, September 2, 1991, p. 70; "Centaur Passes Critical Milestone," *Military Space*, December 16, 1991, p. 3.

[11]U.S. General Accounting Office, *Space Launch: Cost Increases and Schedule Delays in the Air Force's Titan IV Program*, May 1990, pp. 18–19.

The manufacturing of the launch vehicles themselves accounted for only 48 percent of the Titan IV program cost to the Air Force. The remaining 52 percent is for research, development, test, and evaluation (RDT&E); operations and maintenance; and military construction.[12] Thus, this program is not a simple procurement of launch services. The Air Force wants to use the program to develop a heavy-lift vehicle to its specifications. The contract has a target cost, a target price with profit, and a ceiling price. The target and ceiling prices are 10 percent and 20 percent, respectively, higher than the target cost. The Air Force will pay 90 percent, and the contractor only 10 percent, of the cost overrun, up to the ceiling price. The contractor's share will be paid in reduced profit, payable under the contract. These features are similar to the Fixed Price Incentive, Firm Target (FPIF) contract used for MLV-1 and -2. The large number of significant technical modifications and cost adjustments to the contract, however, made the "fixed price" lose its meaning. In addition, a large portion of the contract deals with RDT&E. We therefore consider that the Titan IV program, in essence, is more a cost-plus than a fixed-price contract.

The current version of Titan IV has a lift capability of 39,000 lb to low earth orbits (LEOs). If the Air Force decides to proceed with the solid rocket motor upgrade (SRMU) or to replace it with an upgrade of similar performance, the Titan IV capability could be enhanced to 49,000 lb; it could then deliver most of the Air Force's heavy payloads to LEOs, geosynchronous, and other orbits.

MLV-1. Again on the heels of the Challenger accident in 1986, the Air Force quickly took action to procure expendable launch vehicles (ELVs) other than Titan IVs. The MLV-1 program is intended for launches in the 11,000-lb class to LEOs. The primary payloads are GPS navigational satellites, each one weighing 1800 lb. Their orbits are at semi-geosynchronous altitudes. In 1987, the Air Force

---

[12]U.S. General Accounting Office, *Space Launch: Cost Increases and Schedule Delays in the Air Force's Titan IV Program,* May 1990, pp. 11, 20. As of October 1989, launcher procurement accounted for $4 billion of the $8.3 billion to be funded by the Air Force. There is a small component of $48 million, or .6 percent, for procurement other than launchers. Moreover, the military construction cost of $464 million does not include most of the cost of launch facilities, which are included in other programs. Finally, the other federal users are expected to pay an additional $6.3 billion. The total contract value for the Titan IV program, including 55 launchers, is estimated to be $14.6 billion.

awarded McDonnell Douglas a contract for 20 Delta II launches through 1991 at a market value of $650 million.[13] There is also an option for 24 to 27 additional launches for the 1994–2000 period.

The differences between a government launch contract and a commercial one lie in two areas. There are many Federal Acquisition Regulations (FARs) and other provisions dictated by Congress and other agencies for doing business with the government. These provisions, such as those introduced for socioeconomic reasons, govern almost all types of procurement. Since a reform in this area would affect all government business, not just the launch business, we will not address these issues in this report. We will concentrate on the second area, in which the provisions are used to ensure reliability and control cost in launch contracts and can be changed by selecting a different type of procurement. In other words, procurement type can affect cost because the contractor will have to spend different amounts of manpower on the paperwork and interactions that the government requires in the launch contract.

From the Challenger accident in January 1986 to the MLV-1 award in January 1987, the nation, as well as DoD, was involved heavily in the development of a new space launch policy. The Air Force was well aware of the nation's and the DoD's desire to support the private commercial launch industry. Consequently, the Air Force reduced government-unique provisions and incorporated commercial features in MLV-1. For example, unlike its control of the designs of previous launchers, such as Atlas E, the Air Force allowed McDonnell Douglas to control the basic design of Delta II. The Air Force also relaxed some government quality-control standards in parts, reliability, test and corrective action.[14] These changes are reflected in Contract Data Requirements Lists (CDRLs) and Cost/Schedule Control Systems Criteria (C-SCSC). The former specifies what documents are required to be delivered to the government. The latter ensures that the contractor will use effective internal cost and schedule

---

[13]Congressional Budget Office, "Encouraging Private Investment in Space Activities," February 1991, p. 19. The other government procurement of Delta is NASA's for three launches through 1995 at a contract value of $140 million, and there is an option for up to nine more launches during 1994–2000.

[14]The Medium Launch Vehicle-1 contract between the Air Force and McDonnell Douglas, January 20, 1987.

management control systems and provide timely data that will allow the government to determine contract status. The Air Force was willing to certify a contractor's quality control process and management system rather than inspect and approve numerous intermediate steps throughout the production process. Consequently, over a thousand mandatory inspection points in the old Delta I NASA contract have been reduced by a factor of ten in the Delta II MLV-1 contract. The reduction speeds up the process and reduces the manpower required for compliance.

Government monitoring and contractor compliance can be further reduced. We believe that one source of paperwork is the use of both target and ceiling prices, instead of just one fixed price, in a procurement contract. The MLV-1 contract is a FPIF contract. The contract shows a target cost, a target profit, a target price, and a ceiling price. The target profit is about 10 percent of the target cost, and a ceiling price is 125 percent of the target cost. The target price is simply the sum of target cost and target profit. The structure of the contract requires the government to determine (i) how much the contractor's cost, plus the allowed profit, has exceeded the target price but has not yet exceeded the ceiling price, and (ii) whether the cost, plus profit, has exceeded the ceiling price. The government will pay the lion's share of the cost overrun, up to but not exceeding the ceiling price. Under such a contract provision, the government will have to obtain detailed cost data from the contractor and monitor the contractor's manufacturing to determine actual cost and to control cost overruns. The types of cost data required are, for example, those appearing in work measurements for various jobs and in Cost Performance Reports. As a result of these reporting requirements, the contractor first has to estimate the amount of material and labor required for a job, then document the actual amounts and explain overruns above a certain variance. All these activities take time and manpower. If the contract were a pure fixed-price type and contained a single price, cost data reporting would not be needed, because the government would pay the same price whether there were cost overruns or underruns. Moreover, since the contractor would keep 100 percent of the cost savings and pay 100 percent of the cost overruns, it would have much stronger incentives to keep the cost under control and adopt innovations to reduce costs. Cost reduction procedures developed under government contracts are likely to be applicable to commercial launch customers as well and thus enhance competitiveness. For routine and standard launchers, pure fixed-price contracts could be more suitable than cost-plus contracts or FPIF contracts. For many forms of terrestrial transportation, the government has used fixed-price contracts to procure services. Buying commercial plane tickets for government traveling is one example.

As to mission success incentive (MSI), McDonnell Douglas can collect $3 million for each success.[15] With one failure, it would have to return the collected MSIs and forgo present and future MSIs. With two failures, it would be required, in addition, to reduce the target profit by 50 percent. With three or more failures, it would have to reduce the target profit to zero and eliminate any share of the contractor's award as a result of any underrun. What kind of insurance is the Air Force, in effect, getting?

At an average launch price of $33 million payable to McDonnell Douglas,[16] the target profit is $3.3 million. If there were no failures in 20 launches, McDonnell Douglas would collect $60 million in MSIs and $66 million in profits, or a total of $126 million.[17] If there were one failure, it would only collect $66 million; two failures, $33 million; and three or more failures, $0 million. One failure out of 20 amounts to a reliability of 95 percent. McDonnell Douglas has had a long string of launch successes since September 1986. Delta reliabil-

---

[15]The Medium Launch Vehicle-1 contract between the Air Force and McDonnell Douglas, January 20, 1987, Attachment 6.

[16]Even when one uses the same launcher and configuration, the launch cost can be different from contract to contract. The launch cost to the government can be lower than that to a commercial customer. The difference can be explained by the savings to the government for multiple buys and by the government's paying for a portion of the R&D work through previous contracts. Here, we arrived at the cost per launch by simply dividing the contract value of $650 million by the number of launchers, 20.

[17]There will also be award fees, which are determined by the Air Force and based on the scoring of McDonnell Douglas' management performance in the MLV-1 program. The maximum amount and the actual amount are not shown in the contract description. We can, however, assume that it is the same as that in the MLV-2 program, which stipulates a maximum of $1 million for each of the three evaluation periods. (The Medium Launch Vehicle-2 contract between the Air Force and General Dynamics, June 16, 1988. Attachment 8, p. 3) It suffices to say that these amounts are relatively much less important than the mission success incentives and the target profits. We are ignoring the award fee in the discussion here.

ities are 98 percent in the last 10 years and 94 percent over the last 30 years.[18] Let us conservatively assume that one launch failure is expected. Then, any additional failure has a penalty of $33 million, or the cost of a launch.[19] Therefore, from the Air Force's perspective, the MLV-1 contract is equivalent to insuring the launch vehicle but not the payload, which is about $46 million for a GPS satellite. We will use the finding that MLV-1 contract has insurance on the launcher in Chapter Four.

MLV-2. In 1988, the Air Force awarded General Dynamics a contract for 11 Atlas launches through 1995 at a contract value of $520 million and an option for four additional launches between 1995 and 2000. The primary payloads are DSCS satellites.[20] The contract type is FPIF, the same as MLV-1. It has a target profit and a ceiling price and a specific ratio for sharing the cost overruns and underruns.

Overall, the Air Force considers that it has somewhat less monitoring on MLV-2 than on MLV-1. It intends to maintain its oversight through auditing the contractor's system evaluation. At the beginning of the contract, the Air Force audited and certified the contractor's management system for its ability to produce government specifications and CDRLs. The contract, however, states that

> The Government reserves the right, upon determining that the Contractor's performance in an area subject to this requirement has subsequently become unsatisfactory, to unilaterally reimpose one or more of the Government compli-

---

[18]From June 30, 1981, to June 30, 1991, there were 51 Delta launchers with one failure. From the first Delta launch on May 13, 1960, to June 30, 1991, there had been 205 launches with 12 failures.

[19]Strictly speaking, the insurance covers up to three launch failures. The Air Force receives no compensation or reflight for the fourth and subsequent failures. On the other hand, the probability for four or more failures out of 20 launches (a reliability of 80 percent or less) is very small, because the reliability of Delta launchers well exceeds 90 percent.

[20]Congressional Budget Office, "Encouraging Private Investment in Space Activities," February 1991, p. 19. The other government Atlas contracts are those of NASA and the Navy. NASA procured three launches through 1995 at a contract value of $205 million, and there is an option for two more launches between 1996 and 2000. The Navy procured through Hughes Aircraft Company for one launch in 1991 and options for nine more launches through the 1990s. The total value (including the nine options) of this Navy contract is $700 million.

> ance documents listed below at no increase in contract price,
> or change in the period of performance.[21]

In other words, if the government found that the contractor's management system no longer complied with government requirements, the government could reimpose certain standard documents at no cost. We did not find that the potential cost implication of such a unilateral action was a worry for the contractors. They did not adjust their bid price to account for the provision. They believe that the Air Force is genuinely interested in relying on the contractor's management system and would not unreasonably demand changes to the system. The Air Force also stopped participating in certain review boards that deal with corrective and disposition actions and from certain testing and modeling activities.

A certification and a brief narrative of work accomplished will accompany each progress payment voucher. The time required to prepare such a voucher would be significantly less than the time required to prepare a detailed description of cost and material for payment. The question is whether we can further simplify the documentation by basing the payment only on the passage of time.[22] This way, no work description would be needed, and the government also would not need to check how much work had been performed. For example, 10 percent of the launch price will be paid 36 months before the scheduled launch, and an additional 10 percent will be paid every four months. An interest adjustment or penalty can be made for a launch delay.

The MLV-2 contract simply stipulates a free reflight for failure. In contrast, although MLV-1 essentially demands the same, the Air Force stipulates a complicated procedure to accomplish it. On the other hand, both MLV-1 and -2 do not insure the cost of the satellite.

Although the Air Force has made a concerted effort to incorporate commercial features and approaches into MLV-2, there are still great

---

[21]The Medium Launch Vehicle-2 contract between the Air Force and General Dynamics, June 16, 1988, Part I: The Schedule, p. 70.

[22]The Air Force said that it has since changed the payment method to that based on the passage of time for MLV-2. It said, however, that "the contractor has to demonstrate substantial progress prior to such payments." To us, if the contractor has to demonstrate first, it is not a payment schedule based purely on the passage of time.

differences in required documentation and compliance between MLV-2 and a typical commercial contract. The MLV-2 contract is still three inches thick, whereas a typical commercial contract is less than one-tenth of that. Dennis Dunbar of General Dynamics, the company that won the MLV-2 contract, gave a similar example.[23] In early 1987, General Dynamics received a request from a commercial company and one from the U.S. government. Both were for launch service to deliver spacecraft of similar size to similar orbits. General Dynamics' proposal to the commercial company required 91 pages, but the one to the government, 4250 pages. He further observed that later proposal requests from the same government agency showed significant improvements and that the last request could be answered in 500 pages. Moreover, the bidders were encouraged to propose commercial clauses to replace standard government clauses. Again, General Dynamics' experience is another example that both government RFPs and launch contracts have ample room for streamlining.

There is a price certification clause in the MLV-2 contract that is typical of government contracts. Its purpose is to make sure that, if a contractor charges any other customer at a lower price, the government can also benefit from the same lower price through the existing contract. In the case of space launches, where U.S. contractors are facing competition from heavily subsidized foreign launch providers, such price certification might preclude a contractor from meeting the competition by lowering the price, because a lower price would have to be applied to government contracts and would erode the contractor's profit.

In theory, if a contract, such as that of MLV-2, results from competitive bids, there is no need for such a clause. In a market economy, a merchant has much more freedom to lower the price of his merchandise any time he wants. Government contracts should not restrict a contractor's flexibility.

---

[23]Dennis Dunbar, Vice President of Programs and Technical Operations for General Dynamics Commercial Launch Services, Inc., Testimony of COMSTAC Subcommittee on Procurement to Subcommittee on Space Science and Applications, November 9, 1989.

Before winning the MLV-2 contract, General Dynamics took the risk of investing in the development of Atlas II through upgrading the flight-proven Atlas I.[24] It had also assumed that at least 18 Atlas II and similar versions would be built in the future. In contrast, in the past, General Dynamics had relied on the government to fund launch development from the start. By March 1989, a year after the award of MLV-2, General Dynamics planned to build 60 Atlases through 1997 and have a launch rate of eight per year.[25]

Both General Dynamics and McDonnell Douglas are counting on using similar vehicles to serve the military, civilian, and commercial markets. Also, they count on having business from all three markets. In contrast, Martin Marietta relies much more heavily on the military market for Titan business. The decision might have been greatly influenced by the Air Force's much stronger financial commitment to Titans than to Atlases and Deltas. The Air Force committed $14.6 billion for 55 Titans and only $0.65 billion for Deltas and $0.52 billion for Atlases. Martin might have relatively less incentive to go after commercial business. More important, the Air Force-driven performance requirements make Titan IV, as well as its commercial version, Titan III, oversized for launching commercial satellites one at a time. Titan III can launch 4200 to 5000 lb to geosynchronous orbits. Martin recently decided to stop matching up satellite customers who only require a portion of the Titan III capacity. This decision has placed Martin at a disadvantage with its key competitor, Arianespace. The current generation of launch vehicles, Ariane 4, can use a combination of different solid and liquid strap-on boosters to provide a range of capacities from 2000 to 5000 lb. Arianespace accepts orders of partial payloads and typically matches them to be launched one to three satellites at a time. Arianespace is a tough competitor for Martin, because Arianespace offers a full range of capacity and a very competitive price. When Ariane 5 comes into service in the latter half of this decade, Titan III will fare even worse, because Ariane 5 is expected to offer greater reliability, higher performance (7500 lb to geosynchronous orbit [GSO]) and lower cost than Ariane 4. With the

---

[24]"Atlas Goes Commercial," *Spaceflight*, September 1990, pp. 299–303.

[25]Ibid. General Dynamics plans to stretch the production schedule from year 1997 to 2000. Debra Polsky, "Mission Boosts Atlas Revenues, But Second Write-off Possible," *Space News*, December 16–22, 1991, p. 18.

Air Force providing much of the Titan business, Martin might naturally try to focus its efforts on meeting the Air Force's requirements and compliance standards. Martin's strategy is likely to meet the Air Force's needs first and then garner whatever commercial business it can get. It would be hard for Martin to divert much of its financial and management resources and to deviate significantly from the basic Titan design and practices for the commercial business.

This discussion is not to suggest that the Air Force should reduce its commitment to Titans. After all, the Air Force has its own mission requirements to meet. Rather, the point is that the Air Force's actions have a strong influence on whether Titan can compete in the international market. One path is to have Titan primarily serve the heavy-lift U.S. government market. The other path is, at the Air Force's initiative, to streamline the manufacturing process and to take advantage of the activities related to NASA's various ELV technology programs and the development of the National Launch System (NLS) to make some Titan versions competitive in the commercial market.

Navy

The Navy is more amenable to commercial launches and has used commercial launch procurement for its ten UHF Follow-on satellites. These satellites are of highest priority to the Navy and are urgently needed to replace Fleet Satellite Communications (FLTSATCOM) and LEASAT satellites.[26] A key reason that the Navy decided on commercial launch procurement is that it had a rather painful experience with government launches. NASA was in charge of the launches of eight FLTSATCOM satellites. Under NASA's supervision and control, there were two launch failures out of eight, for a reliability record of merely 75 percent. The Navy estimated that these accidents cost over $400 million. On the other hand, the Navy had a fine experience with the LEASAT program, which simply leases communication services after satellites are successfully operating in orbit. The Navy's decision for a commercial launch procurement

---

[26]"Navy Satellites Approach Critical Replacement Stage," *Aviation Week & Space Technology*, March 21, 1988, p. 50. LEASATs have been used to provide the Navy with a leased satellite relay capability.

was also influenced by the recent DoD and national space directives that encouraged the use of commercial launch procurement.

Since the government emphasizes holding program managers responsible for the outcomes of programs, one way to enhance the chance for success is for the program manager to structure the procurement contract so that the contractor has strong incentives to deliver payloads to orbit on time and within budget. In the UHF Follow-on contract, for example, the contract stipulates free reflights. Moreover, if a payload is lost, the Navy will be reimbursed 80 percent of the satellite cost. There are delay penalties on the first two flights. The Navy probably believes that these provisions will give Hughes stronger incentives than a traditional government contract or even than the more recent commercial-like contract in ensuring mission successes.

There has been continuing debate about whether the commercial launches of UHF Follow-on satellites will save the Navy money. In a report on the National Defense Authorization Act for FY 1990 and 1991, dated July 19, 1989, the Senate Armed Services Committee directed the Under Secretary of Defense for Acquisition to reevaluate the Navy's plans for commercial-type launches of UHF Follow-ons. In the response, the Air Force was stated to have estimated that procuring launch services through the Air Force's MLV-2 program could save $120 million if there were no launch failures or $78 million for one failure. The Navy disagreed. With no failure, the Navy estimated in April 1990 that commercial procurement would have saved $92 million out of a total launch cost of $801 million, or 11 percent.[27] With the launch reliability record of Atlas at about 87 percent, it is much more likely to have one failure than none. Based on the Navy data, we estimated that the savings of using commercial procurement could be $135 million, or 16 percent, for ten launches with one failure. The controversy of whether commercial launch procurement saves money persists. General Robert Dickman, Deputy Director of Air Force space acquisition, told the House Science Space Subcommittee that if the Atlas continues to be as reliable

---

[27]Report delivered to the Committee on Armed Services by Donald J. Yockey, Principal Deputy, the Office of the Under Secretary of Defense, on April 6, 1990.

as it has proved over time, the Navy would have saved money by buying Atlas launch services through the Air Force.[28]

There are two other reasons that the Navy is more amenable to commercial launch procurement. First, the UHF Follow-on satellite was estimated during 1989 to be only $54 million, relative to a launch cost of about $71 million.[29] In contrast, Air Force satellites typically cost more. A Defense Support Program (DSP) satellite would cost $221 million, while a DSCS III would cost $92 million.[30] All else being equal, a lower satellite cost allows one to be more willing to assume a higher risk in trying out the newer commercial launch procurement. Second, UHF Follow-on satellites are replacements of FLTSATCOM. If a failure occurred, the delay would be manageable, because some FLTSATCOM satellites will likely remain in orbit. Also, the planned launch schedule of ten UHF Follow-ons in a short period of five years allows at least some services, albeit at a lower level. Each of these UHF Follow-on satellites will have a minimum life expectancy of 10 years and a mean life of 14 years.[31] In other words, there is a danger of degraded capacity but little danger of a complete service void over a certain area. In contrast, Air Force satellites, such as DSP, have less redundancy, and the possibility of a gap in coverage is a very serious concern. The Air Force is justifiably more reluctant to try newer procurement methods that reduce their control and oversight.

Strategic Defense Initiative Organization

SDIO has used commercial launches for some of its payloads. It procured a Delta II commercially for the launch of LACE (Low-Power Atmospheric Compensation Experiment) and RME (Relay Mirror Experiment). This reflects SDIO's faith in commercial launch pro-

---

[28]"Air Force rejects commercial launches," *Military Space*, July 29, 1991, p. 6.

[29]As of June, 1990, the average price of a UHF Follow-on satellite remains slightly under $60 million. The cost of the overall UHF program is slightly under $1.4 billion, with satellites and launches sharing the costs about equally. Michael A. Dornheim, "Navy Likely to Add New Capability to UHF Follow-on Communications Satellites," *Aviation Week & Space Technology*, June 4, 1990, p. 69.

[30]C. L. Whitehair, "Costs of Space and Launch Systems," The Aerospace Corporation, February 20, 1990.

[31]"Navy Satellites Approach Critical Replacement Stage," *Aviation Week & Space Technology*, March 21, 1988, p. 50.

curement, because LACE and RME have a combined price tag of $275 million relative to a launch cost of merely $38 million. Colonel Thomas W. Meyer, Director of the SDIO Directed Energy Office, was quoted as saying that the commercial procurement has resulted in an estimated savings of $13 to $17 million,[32] a sizable cost reduction of 25 percent to 31 percent.

On the other hand, the SDIO procured the prior three Delta vehicles for the Delta 180 series of launches through the Air Force. They were government launches, because significant modifications to the second stage were needed. SDIO's launch policy, according to Colonel Gary E. Payton, SDIO Associate Director for Technology, is to use commercial procurement if a launch is straightforward and to use government procurement if substantial modifications are necessary. We agree that the degree of launcher modification is an important determinant in selecting commercial or government launch procurement. Finally, like the Navy, SDIO is strongly influenced by the directives to use commercial launch services whenever possible.

DARPA

DARPA has been the leading sponsor of light satellites and small launch vehicles. Orbital Sciences Corporation's Pegasus and Taurus launchers have benefited greatly from DARPA's support. Pegasus has used commercial practices exclusively, and the same is expected of Taurus. Whereas there are widely differing opinions about what launchers for which payloads should be commercialized, there is a consensus that small launchers are most suitable for commercialization. Moreover, small DARPA payloads are particularly suitable for commercial launch procurement, because experimental or developmental payloads are generally less time critical and often less expensive. DARPA's Pegasus launches were transferred to the Air Force in 1991 upon the completion of the demonstration phase. Moreover, the Air Force has recently awarded its own AFSLV (Air Force Small Launch Vehicle) contract to Orbital Sciences Corporation. The contract, however, consists of only one firm launch and 39 optional

---

[32]Edward H. Kolcum, "SDI Laser Test Satellites Placed in Precise Orbits," *Aviation Week & Space Technology*, February 19, 1990, p. 25.

launches. DARPA's Taurus launches are also expected to be transferred to the Air Force.

Army

The Army's space policy has been to rely on national satellites procured by the other services, rather than procuring satellites dedicated for its own use. The Army uses such satellites as DSCS, GPS, DSP, and the Defense Meteorological Satellite Program (DMSP) as force multipliers for its AirLand battle-fighting doctrine. Our study of various Air Force launch and satellite programs will automatically cover those satellites used by the Army. In this report, we do not evaluate Army launch needs separately. Such needs might, however, arise in the future. There are studies that recommend that the Army obtain satellites of its own for such missions as over-the-horizon surveillance or in-theater mobile communications.

Department of Commerce

Among government agencies, the Department of Commerce (DoC) is the most concerned about the competitiveness of the U.S. launch industry in the international market. The DoC is naturally concerned that the low launch prices offered, especially by such nonmarket providers as the PRC and the CIS, have unfairly wrested business from U.S. launch providers. On the other hand, banning or limiting U.S. satellites to be delivered by these foreign, heavily subsidized launchers is not a long-term solution. Foreign satellite makers can undercut us by offering package deals that use cheap launchers. The DoC is particularly interested in ways to lower the cost structure in the U.S. launch industry. Since the U.S. government is the major customer, many of its oversight and documentation requirements are incorporated into the industry's manufacturing and management process. For example, if a government launch contract demands the right to approve the manufacture of a launch vehicle at various checkpoints, a contractor would design the manufacturing process and the internal data management system to accommodate such requirements. Since the same type of launcher is generally used for both government and commercial payloads, a contractor will use the same process and system for both, because a launcher is not designated for a specific payload until near the very end of production. Therefore, if the government requirements increased the manufacturing cost for government buys, they would increase the cost for commercial buys as well. The DoC does not want government requirements to increase the launch cost for commercial payloads.

Department of Transportation

The Commercial Space Launch Act of 1984 designated the Department of Transportation (DoT) "as the lead Federal agency to facilitate and expedite the issuance and transfer of commercial space launch licenses." The Act also requires "any person engaged in non-Government launch operations in the United States to obtain a license." Consequently, DoT is most concerned about which launches are considered government launches and which are nongovernment. Only the latter category is under the jurisdiction of DoT. According to our launch classification, both government and commercial-like launches are officially "government launches," and only commercial launches are "nongovernment." The DoT has expended great effort in establishing regulations for nongovernment launches or, as designated in this report, commercial launches. DoT is striking a balance between ensuring public safety and avoiding burdensome regulations.

The DoT generally treats space launch as a mode of transportation that is not fundamentally different from aircraft, motor vehicles, trains, or ships. In a major study, the Licensing Programs Division of the Office of Commercial Space Transportation concluded that "while there may be some specific military payloads which for reasons of their necessity to national security, need for special hardware support, or requirements of coordination with other nations, are appropriately launched by the military itself, the majority of military requirements can be fully met by commercial firms."[33]

---

[33]Licensing Programs Division, Office of Commercial Space Transportation, Department of Transportation, *National Space Transportation Infrastructure: The Commercial Impact*, August 1988, p. 2.

U.S. Launch Contractors

Traditionally, NASA and the Air Force have purchased launch vehicles from the U.S. launch industry and controlled their launches. In fact, NASA and the Air Force have also controlled many aspects of vehicle design and manufacturing through specification, inspection, milestones, and other types of review and oversight. The first commercial launch or launch under the control of the contractor occurred in August 1989. Although the launch industry has in recent years invested over $600 million of its own funds in improvements of launch manufacturing and processing facilities, much of its manufacturing infrastructure was funded by the government, and it continues to use Air Force launch sites, facilities, and personnel for launch processing. It also found some of the government oversight useful in providing an independent check. More important, it depends on the government for the bulk of the launch business. These considerations make industry an unlikely candidate to do things that would displease the government clients. The best we can hope for is that when industry finds areas in which the government could relax documentation and reporting requirements, it would make them known to the government. It would be up to the government to take the initiative for streamlining action. Through the Commercial Space Transportation Advisory Committee (COMSTAC), the launch industry has been making its recommendations known to the Secretary of Transportation and other launch planners.

Chapter Four

A MODEL FOR SELECTING A LAUNCH
PROCUREMENT TYPE

INTRODUCTION

Many factors determine how a DoD launch contract should be procured—by GL, CLL, or CL. In this chapter, we use a model to capture the monetary tradeoff of two key factors: the potential cost savings and possible lower reliability in using a particular type of procurement. Chapter Five will discuss those key factors that are difficult to measure in monetary terms, including the extent of contract modifications during contract performance, the likelihood and degree of mission delay if a particular procurement type is used, and the feasibility of switching back to a more closely monitored contract if too many launches fail. We will develop the model and then apply it to four existing launch contracts and one upcoming one: MLV-1 for GPS launches, MLV-2 for DSCS-III launches, Titan IV program for DSP-BL IV launches, the Navy/Hughes contract for UHF-FO (UHF Follow-on) and the upcoming MLV-3 for GPS-FO. We want to see whether our model indicates that the existing contracts should have been procured differently—for example, commercial launches instead of government or commercial-like launches. We also want to know what type of procurement contract is recommended by our model for the upcoming MLV-3.

MODEL OBJECTIVES

Much of the current debate on whether DoD should use commercial launches hinges on their reliability. Commercialization advocates argue that commercial launches are as reliable as government launches. The opponents worry that commercial launches either are less reliable or have too small a sample to provide high confidence that they are as reliable. The opponents further argue that they cannot afford to find out.

The problem stems from many program managers' unwillingness, with good justification, to trade a much longer reliability track record in government launches for cost savings in commercial launches. We think that DoD might consider a different approach, namely, adding some features to commercial launches so that a payload's mission performance success depends less on individual launch success. The strategy is to see whether commercial launches can be designed for certain DoD payloads so that commercial launches are more comparable to government launches in every important way—cost, timeliness of payload's mission performance, and the like. In other words, a program manager does not have to do much trading off, something that he does not want to do in the first place. If such commercial launches can be designed and DoD payloads can be found, we can commercialize those launches first.

A program manager's worries in going commercial are mostly that the actual commercial launch record could turn out to be worse than the current record, which consists of only a small sample, and that the current commercial reliability record is inapplicable to his "unique" payloads. He is concerned about (i) possible mission delay and (ii) the potential monetary loss of the payload and/or the launcher. We believe that the former can be reduced by having a backup satellite and launcher at the launch site and that the latter can be alleviated by purchasing insurance on both the satellite and launcher.

As discussed above, our model is based on the premise that one way to convince a program manager to switch from government or commercial-like launches to commercial launches is to show that he will save money without added worry of degraded or delayed mission performance. We assume that, if commercial launches are chosen, the procurement contract will include the cost and provisions to support a backup satellite and launcher (Table 4.1).

The backup can be useful in three situations. First, if there is a delay in the manufacturing of launchers and satellites, the backup can reduce a delay in the launch schedule. Second, if a launch fails, the

Table 4.1

Launch Selection Model

- Objective: same number of satellites in orbit
- Assumes CL would have
  — Lower cost per launch
  — Lower reliability
    - Compensated by additional launches
    - Compensated by a backup launcher and satellite at launch site
    - Compensated by an insurance option to ensure funds for replacement

> How low can CL's reliability be
> before wiping out its savings?

backup can be used immediately after the standdown.[1] The standdown times for ELVs range from 2.8 months to 5.2 months.[2] Without a backup, it might take longer to catch up on the satellite deployment schedule. Third, a satellite manager's key worry is that should a crisis or war develop during a standdown, the satellite would not be in place when it is needed the most. In such a circumstance, a backup might be considered for emergency launch even before the post-failure investigation is fully completed. The time to process and integrate the backup satellite and launcher will be about a couple of months. Such a launch is especially justified in cases where preliminary investigation provides high confidence that the cause of failure is not generic or can be fixed quickly in the backup.[3]

---

[1]After an in-flight failure, the current practice is to stop launching vehicles with a similar design or that use the same suspected failed equipment. The failure is investigated, and design or procedural changes are recommended, implemented, tested, and validated. The time from failure to resumption of launch is called the standdown time.

[2]The data were through April 3, 1987. The average standdown time for Delta in 12 failures since 1960 was 2.8 months. That of Atlas/Centaur was 4.1 months in ten failures since 1962. Titan IIIs had 5.2 months in ten failures since 1963. In contrast, the shuttle had a much longer standdown time, over 24 months, for its single failure. Harry Bernstein, "Space Launch Systems Resiliency," the Aerospace Corporation, *Proceedings of the Twenty-Fifth Space Congress*, Cocoa Beach, Florida, April 26–29, 1988.

[3]Another option, on-orbit spares, is not discussed in this report because the option is the same for all three types of launch procurement.

Take the example of launching 20 GPS satellites. The existing MLV-1 has a commercial-like launch contract. Had the Air Force procured MLV-1 commercially, we would assume that, even during the first launch, an additional satellite and launcher would be available as a backup. One does not need, however, to buy an extra satellite and launcher in the program for the backup; one only has to buy the additional pair early. If the very first launch is delayed or fails, the backup could be used in one of the three ways described above. If the first launch succeeds, the backup could be used for the next scheduled launch, and the newly produced satellite and launcher could be used as backup for the second launch. This rotational scheme prevents the backup from becoming obsolete and avoids the need for procuring an extra satellite and launcher. As for cost implications, this backup scheme is equivalent to the purchase of the last satellite and launcher in the program at the same time as purchase of the first satellite and launcher. There will be an additional cost for maintaining the backup throughout the period that launches are taking place.[4] Finally, our backup scheme makes the model unsuitable for the analysis of a launch program consisting of only one satellite.

An early purchase is more costly because of the time value of money, and this added cost has been factored into our model. While standby is an attractive concept for better system reliability and timeliness regardless of launch procurement type, the key point is not that a commercial launch have a standby but that provisions have been made for it.

MODEL DESCRIPTION

We are now ready to develop a model that will allow us to carry out a break-even analysis between any two of the three procurement types: GL, CLL, and CL. For ease of presentation, our formulas compare commercial launch procurement (signified by letters cl) and government procurement (signified by letters gl). We will use the same formulas for comparing commercial procurement with commercial-like procurement, and will simply substitute the parameters with letters gl with those for the commercial-like procurement (cll). Figure 4.1 and Table 4.2 show the approach and parameters to be

---

[4]We will indicate that this cost is relatively insignificant.

used in the model. The break-even condition is the changeover point at which the government's better procurement strategy changes from one type of procurement (GL or CLL) to the other (CL).[5] Since the condition depends on whether the government mandates the launch contract's inclusion of insurance on launchers and/or satellites, we define three categories: (1) no insurance, (2) insured launcher, and (3) insured launcher/satellite. These three categories can be used in two ways. First, if the government decides on a certain insurance policy, it can use the model to determine the best procurement method. Second, as we prefer, the government uses all three categories to help decide on both the most suitable insurance policy and the most effective launch procurement method.

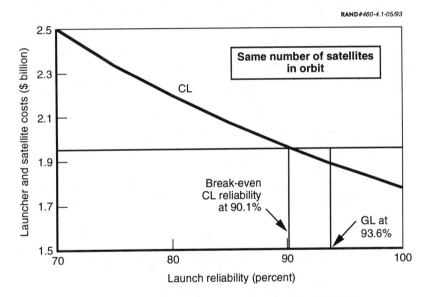

Figure 4.1—Determination of Break-Even CL Reliability

---

[5]In the model, we group CLL with GL because the main cases of study are GL versus CL and CLL versus CL, not GL versus CLL. It should be emphasized that the model can also be used to analyze the case of GL versus CLL by simply replacing CL parameters with CLL ones in the formula.

Table 4.2

Parameters in the Procurement Selection Model

S	=	satellite cost in million $
Lgl Lcll Lcl	=	cost, excluding insurance premium and before adjustment for launch failures, for each government, commercial-like, or commercial launch in million $
N	=	number of satellites that one wants to place in orbit
Rgl Rccl Rcl	=	reliability of government, commercial-like, or commercial launches in percent
T	=	time between first launch and last launch in years
Cbu	=	cost of maintaining the backup satellite and launcher throughout the period T
EIFR	=	extra insurance fee rate = total premium rate - expected launch failure rate, all in percent
FIC	=	government's investigation cost of launch failure
r	=	discount rate in percent
Ft	=	present value factor to account for the time value of money in a stream of equal payments
UCSLcl,gl UCSLcl,cll	=	savings per launch before adjustment for launch failures = $(1 - Lcl/Lgl)*100$ or $(1 - Lcl/Lcll)*100$ in percent
DR	=	difference in reliabilities, $(Rgl - Rcl)$ or $(Rcll - Rcl)$
Ngl Ncll Ncl	=	number of government, commercial-like, or commercial launches = $100*N/Rgl$, etc.
Igl Icl Igs Ics	=	denotes whether insurance is purchased for launchers or satellites. For example, $Igl = 0$ if no insurance on launcher is included in a GL contract. If insurance is purchased, $Igl = EIFR$

In category one, the government does not intend to purchase insurance on launchers and satellites.[6] This is the traditional way for the government to procure launch services. The government reasons that it has enough projects and financial resources to be self-insured and thus can save by not paying fees to a profit-making insurance company. Unfortunately, what the government saves in fees could be overwhelmed by the cost of oversupervision. How high are the fees? To begin, we must discuss what launch insurance covers. It generally covers two types of accidents: launch failure and satellite initial on-orbit checkout failure. The average premium was stable at about 10 percent of the insured value (launcher and/or satellite) from 1976 to 1979 and varied considerably thereafter.[7] It averaged 12 percent in 1981 and 6 percent in 1982. From 1983 to 1984, it fluctuated between 5 percent and 10 percent. Then, as a consequence of increasing launch failure rates, the premium rose almost monotonically to the 25 percent level by 1987.[8] As of October 1991, the premium rates stood in the 17 to 20 percent range.[9] The probability of failure during initial satellite checkout is attributable to the manufacturing of the satellite.

Since the current study deals with launch, but not satellite, procurement types, we need to know the portion of the premium that is for the coverage of launch failure alone. Unfortunately, the premium is seldom broken down in separate rates for launch failure and satellite checkout failure. Our discussions with various issuers and users of insurance indicate that launch failure coverage accounts for somewhat more than half of the premium rate, or somewhat higher than 8.5 to 10 percent. The precise rate will depend on the specific launch record of the launcher type under consideration and many other factors. Based on recent insurance rates charged by the insurance industry on Deltas, Atlases, and Titans, we deduced that the premium rate for launch failure alone is about 4 percentage points

---

[6]In this report, insurance refers to launch insurance against the loss of launchers and/or satellites during launch and does not include satellite on-orbit checkout insurance.

[7]This is not the third-party insurance that covers the damages to third parties such as the public and the launch site facilities.

[8]Giovanni Gobbo, "An Insurer's View of the Space Business," *Space Policy*, February 1991, pp. 47–49.

[9]Daniel Marcus, "Record Insurance Deal Sought," *Space News*, October 7–13, 1991, pp. 3, 21.

above the expected launch failure rate.[10] Thus, we use an extra insurance fee rate of 4 percent (see Tables 4.1 and 4.3) for the calculations in this report.[11] At issue is whether insurance is a substitute for government oversight to ensure launch successes and whether insurance can reduce oversight cost by more than the EIFR. As will be discussed in the section on Values of Unadjusted Cost Swings Per Launch, if commercial launch procurement can reduce launch cost by 11–28 percent, the total cost savings can be substantially higher than the insurance fees of 4 percentage points on launchers and/or satellites.

In all three insurance categories, we assume that, for commercial launches, a satellite/launcher backup will be stored at the launch site. However, the backup is assumed not to be needed for the government and commercial-like launches under comparison. The model can be generalized by assuming that government and commercial-like procurements could also have backups, but that there would be an additional satellite/launcher backup under commercial procurement. Since both assumptions should give similar results, we will not provide the formulas for the generalized model.

BREAK-EVEN POINTS

The break-even point in the model is determined by equating the total costs of two launch procurement alternatives. For example,

Cost of government launch = Cost of commercial launch

---

[10]The premium rate is the sum of two components: the expected launch failure rate and the extra insurance fee rate. The former is different for different launchers. The latter is charged by a private company for the cost of doing business and profit and is absent if the government is self-insured.

[11]In principle, the extra insurance fee rate could be lower when a launch contractor self-insures, and we can run our model with a lower EIFR for that case. In practice, launch contractors want to hedge against this kind of risk, for the same reason that international companies go to external markets to hedge against exchange rate risks or raw material price fluctuations. Therefore, even if contractors self-insure, it is unlikely that they would demand an EIFR considerably lower than what an outside insurer would charge. Another reason for self-insurance is when contractors cannot obtain outside insurance for launches, say, more than three years away.

$$\overbrace{(100+Igl)\frac{N}{Rgl}Lgl*Ft}^{\text{Launch cost}}+\overbrace{(100+Igs)\frac{N}{Rgl}S*Ft}^{\text{Satellite cost}}+\overbrace{(100-Rgl)\frac{N}{Rgl}FIC*Ft}^{\text{Cost of failure}}=$$

$$\overbrace{(100+Icl)\frac{N}{Rcl}Lcl*Ft}^{\text{Launch cost}}+\overbrace{(100+Ics)\frac{N}{Rcl}S*Ft}^{\text{Satellite cost}}+\overbrace{(100-Rcl)\frac{N}{Rcl}FIC*Ft}^{\text{Cost of failure}}$$

$$\overbrace{(S+Lcl)\left[1-\frac{1}{(1+r)^T}\right]}^{\text{Backup cost}}+\overbrace{Cbu*Ft}^{\substack{\text{Cost of}\\\text{maintaining}\\\text{backup}}}$$

where

$$Ft=\frac{1}{T+1}*\frac{1-\dfrac{1}{(1+r)^{T+1}}}{1-\dfrac{1}{1+r}}$$

Let us examine what each term represents. The terms on the left-hand side are the cost components of government launches. The first term is the expected total launch cost in successfully placing N satellites into orbit. The number of launches to yield N successful launches will be 100*N/Rg and the cost is 100(N/Rg)Lg. We have used 100(N/Rg)Lg, instead of 100(N)Lg, in order to include the cost of the expected number of launch failures. Instead of self-insurance, if launcher insurance is purchased from a private insurer, we should expect that the insurer will charge an extra fee, which is above the already included expected launch loss, to cover administrative costs and to earn a profit. The factor is represented by Igl, which is equal to EIFR. If no insurance is purchased, one simply sets Igl to zero. The launch cost is assumed to be paid in equal installments starting at the first launch and ending at the last launch. The factor, Ft, accounts for the time value of money for these installments and brings the discounted cost to the reference time, that of first launch.

Similarly, the second term on the left-hand side represents the total satellite cost for placing N satellites in orbit. Again, if insurance on the satellite is purchased, Igs should be set to EIFR; otherwise, zero.

The third term represents the cost to the government in investigating launch failures.

The first three terms on the right-hand side have the same interpretation as those on the left, except they represent commercial launches, as opposed to government launches. The fourth term denotes the cost of holding a satellite and launcher as backup. The last term is the cost of maintaining the backup throughout the period T.

Let us now review cases with and without insurance. The four parameters, Igl, Igs, Icl, and Ics, denote whether insurance is purchased on launchers (l) and/or satellites (s) in government launch procurement (g) and/or commercial launch procurement (c). For example, if both launchers and satellites are insured in a commercial launch contract but not in the comparative government contract, Icl and Ics will be equal to the extra insurance fee rate, EIFR, and Igl and Igs will be zero. Looking at the break-even equation somewhat differently, we have assumed that a reserve has been established to cover the uninsured, expected losses as a result of launch failures. In the insured case, the reserve is simply used for a part of the insurance premium. The difference between the insured and uninsured cases is that in the former one has to pay an extra insurance fee to cover the insurance cost of doing business and profit.

The equation can be rearranged to yield a decision rule in terms of difference in reliabilities,

$$DR = Rgl\left[1 - \frac{\left(100 + Icl + 100\dfrac{FIC}{Lcl}\right)Lcl + (100 + Ics)\,S}{\left(100 + Igl + 100\dfrac{FIC}{Lgl}\right)Lgl + (100 + Igs)S - \left[(S + Lcl)f + \dfrac{Cbu}{N}\right]Rgl}\right]$$

where

$$f = \frac{1 - \dfrac{1}{(1+r)^T}}{N * Ft}$$

In other words, if the reliability of government launches were more than DR percentage points above that of commercial launches, government launch procurement would be less expensive. If it were less than DR, commercial procurements would cost less.

MODEL APPLICATION TO EXISTING PROGRAMS

The model is now to be applied to four existing cases and one upcoming case, as shown in Table 4.3. In this section, we will analyze whether there would have been any cost savings if the launch contracts for the four cases had been procured differently. We will also analyze the potential cost savings in procuring MLV-3 commercially. In the next chapter, we will study the nonmonetary advantages and disadvantages of procuring MLV-3 commercially.

Table 4.3 lists the five study cases, and Table 4.4 gives the input parameters. The satellite and launch costs are based on those of C. L. Whitehair.[12] The reliability values were based on the records of the

---

[12]C. L. Whitehair, "Costs of Space and Launch Systems," The Aerospace Corporation, February 20, 1990. We did not, however, use his number of $25.2 million for a GPS-Block-IIR satellite, because we do not feel confident that the cost of a GPS satellite can be reduced so drastically. Instead, we use $46 million, which is the cost of the current GPS-Block-II. If any cost reduction materializes, it would reinforce our conclusion that GPS Follow-on is the most suitable for launch commercialization of the five programs in Table 4.2. Moreover, Whitehair's launch cost can be considerably higher

Table 4.3

Five Study Cases for Launch Procurement

Satellite	User	Contract	Year Awarded	Launcher	Procurement Type
DSP-BL IV	Air Force	Titan IV	1985–1989	Titan IV	GL
GPS	Air Force	MLV-1	1987	Delta II	CLL
DSCS-III	Air Force	MLV-2	1988	Atlas II	CLL
UHF-FO	Navy	Hughes/GD	1988	Atlas II	CL
GPS-FO	Air Force	MLV-3	1992	Undecided	CLL in RFP

Table 4.4

Input Parameters of Five Study Cases for Launch Procurement

Parameters	DSP-BL IV	GPS	DSCS-III	UHF-FO	GPS-FO
Contractor liability	Neither	Launcher	Launcher	Both	Undecided
S, $mil	221	46	84	54	46
Lgl or Lcll	218	59	61	61	59
UCSLcll,gl or UCSLcl,cll, percent	0–20	0–28	0–28	0–28	0–28
N	5	20	7	10	20
Rgl or Rcll, percent	93.6	94.3	87.2	87.2	95.4
T, yrs	4	4	4	5	5
r, percent	10	10	10	10	10
EIFR, percent	4	4	4	4	4
FIC, $mil	10	5	7	7	5

SOURCE: The reliabilities data are explained in footnote 13. Satellite and launcher costs were based on C. L. Whitehair, "Costs of Space and Launch Systems," The Aerospace Corporation, February 20, 1990.

relevant launches since 1970.[13] We use a discount rate of 10 percent as reference. In the section on Model Application to the MLV-3

---

than the price paid to a launch contractor, because in addition to launcher manufacturing and payload integration costs, Whitehair's cost includes those of launch operations and government furnished support.

[13]We used three sources of data for launch reliabilities (Karen Poniatowski, "Expendable Launch Vehicle Capabilities, Constraints, and Costs," NASA Office of Space Flight, March 9, 1989; Steven J. Isakowitz, *International Reference Guide to Space Launch Systems*, 1991 Edition, American Institute of Aeronautics and Astronautics; and Delta and Atlas/Centaur data compiled by McDonnell Douglas and General Dynamics, respectively). All these data are consistent except those for Delta. Poniatowski gave 100 successes out of a total of 109 launches for the 1970–1988 period,

Contract, we will show that our results and conclusions are insensitive to the discount rate, because the launch performance period is short and at most five years.

We have not included values for Cbu, the cost of maintaining the backup throughout the period when the satellites in the program are being launched. The maintenance cost depends on the required state of readiness. We expect, however, the cost to be relatively insignificant if the backup is not maintained on pad.[14]

VALUES OF UNADJUSTED COST SAVINGS PER LAUNCH

A key parameter in determining the attractiveness of commercial launches is the unadjusted cost savings per launch, UCSLa,b. This is the savings in using a, instead of b, type of launch procurement. For example, a and b can be commercial launches and government launches, respectively. The UCSLcl,gl is the savings per launch before adjustment for possible launch failure, backup launcher/satellite cost, and insurance cost. The cost savings may come from lower labor costs associated with cost and technical data reporting in

---

while Isakowitz and McDonnell Douglas both gave 102 successes out of 108. Our examination revealed that the latter is correct.

We calculated the reliability figure for each launcher type from 1970 to the end of the year preceding the year in which the launch contract was awarded. We will use the data to examine, in retrospect, whether it would have been more advantageous to procure a launch contract differently. Based on the contract award years shown in Table 4.3, we used the cutoff years of 1986, 1987, and 1987 for MLV-1, MLV-2, and Hughes/General Dynamics (GD) UHF-FO, respectively. For Titan IVs, the Air Force had made initial and additional purchases from 1985 to 1989. Since our interest is in the launches of DSP-BL IV #18 to #22, we use a cutoff year of 1988. For the upcoming MLV-3, the winning contractor is unknown, so we used the Delta figure of 95.4 percent for the period from 1970 to June 30, 1991, as a proxy. In the section on Model Application to the MLV-3 Contract, we will show that our analysis and conclusions for the case of MLV-3 are not sensitive to the precise value of reliability, because we do not compare the reliabilities of various launchers but rather the different types of procurement, even if the same launchers were used.

[14]The cost would be high if the backup is kept on pad for very quick response, say, launchable within a few days. If DoD requires such on-pad readiness for vehicles in the Delta II, Atlas II, and Titan IV classes, these vehicles will form a special type of highly responsive launchers, and their procurement should be treated separately from those covered in the present report, which deals with routine military, civil, and commercial launches.

a commercial, as opposed to a government, launch contract.[15] If the cost to manufacture a launcher and to process a launch under a government launch contract is $100 million and that under a commercial launch contract is $80 million, the UCSLcl,gl would be 20 percent. If the commercial launch reliability is lower and a backup is used in commercial launches, the adjusted or net savings per launch would be less or could even be negative. A negative value means that the commercial launches actually cost more than government launches, after all costs are taken into account.

There are three ways to determine UCSL. First, one may think that one can simply observe the prices charged under GL, CLL, and CL launch contracts by the same launch providers for similar payloads. All three major launch providers—Martin Marietta, General Dynamics, and McDonnell Douglas—use the same or similar launchers in at least two of the three procurement contracts. Unfortunately, it is difficult to derive meaningful price differentials in this way. The number of launches in a GL or CLL contract is generally much higher than that in a CL contract. The adjustment due to bulk buy is unknown. On the other hand, government launch buyers often demand a price certification that requires, if a lower price is charged to commercial customers, that the government will receive the same price also. Thus, launch providers are reluctant to charge a commercial customer a lower price for fear of lowering the profit margin on government contracts, even if commercial launches are less costly to supply. What makes this way infeasible, however, is that the observed price for CLs has already been affected by GLs or CLLs. Recall that contractors use the same production line and monitoring procedure for all launchers, because at the time of production they often do not know whether a particular launcher is used for CL, CLL, or GL. Thus, if the price for GL or CLL is high, the price for CL would also be high. What we are really interested in is the CL price if all launchers are manufactured and processed under commercial launch contracts, not some under CLs and some under GLs and CLLs. Since all three major launch providers serve GLs or CLLs, we cannot observe the pure CL prices in the market place.

---

[15]We are counting only the savings in commercial launches realized by the contractor. We are not including the potential savings by the government in reducing personnel monitoring the launcher manufacturing and processing. This staff may have to be kept if the Air Force wants to switch out of a commercial launch contract quickly in the event that commercial launches fare far worse than expected.

The second way is to analyze engineering and costs to determine the price differential. One can start with the launch price under the current practice of mixed CL, CLL, and/or GL launch production. Then, one can estimate what production and reporting steps and costs could have been saved if all launch contracts were CL. All three major launch providers are able to perform this analysis, and they can have a good idea what that savings will be. They will, however, be willing to reveal the amount of savings in public only if their major customer, the Air Force, gives them its blessing. Otherwise, it is natural for the contractors to be leery of displeasing their major customer. We would recommend that the Air Force encourage and even sponsor such analyses.

The third way is to treat CL, CLL, and GL as options in the request-for-proposal (RFP) and to request that the bidders provide prices for CL, CLL, and/or GL. Price data obtained this way are most relevant because all three procurements are for exactly the same launch program. This is how we decide whether to buy or lease a car; we first get quotes for both. Why should the government ask for quotes for only GL, CLL, or CL? We recommend that at least two of three options (GL, CLL, and CL) be used in RFPs.

Although a comprehensive data base is lacking, some data are available. In congressional testimony, Dennis Dunbar, Vice President of Programs and Technical Operations for General Dynamics Commercial Launch Services, said that the Air Force MLV-2 RFP called for specific tailoring to accommodate commercial features. When compared with General Dynamics' last contract under full-up Federal Acquisition Regulations and untailored military specifications, General Dynamics' specific tailoring of the government specifications resulted in a 64 percent reduction in launch cost; 20 percent is attributed to its commercial approach.[16] This indicates that going from GL to CLL could save 20 percent. On the other hand, the Air Force believes that the savings are considerably less. We use a range from 0 to 20 percent for UCSLcll,gl in our analysis.

There are also some data for the savings in switching from CLL to CL, UCSLcl,cll. In the procurement for launching the LACE/RME satellite, SDIO obtained bid prices for both CLL and CL. The average

---

[16]Testimony of COMSTAC Subcommittee on Procurement to Subcommittee on Space Science and Applications, November 9, 1989.

savings in CL, UCSLcl,cll, were 28 percent.[17] For the Navy's UHF-FO procurement, there were data for both CLL and CL. Based on the Navy's estimate and our assumption of one expected launch failure out of ten, we arrived at a UCSLcl,cll of 16 percent.[18] We should emphasize that both savings were estimated under the assumption that the launch provider still has to produce and launch launchers to meet both types of contracts—CLL and CL. During production, the contractor does not know whether the launcher is for CLL or CL, or does not want to use two different procedures and bookkeeping. Consequently, some CLL costs stay with CL. The CL costs would be lower or the savings larger than 16 percent to 28 percent if a launch provider does not have to serve CLLs at all. On the other hand, for both the SDIO and Navy procurements, the Air Force maintains that savings do not exist. We use a range from 0 to 28 percent for UCSLcl,cll in our analysis.

The cost difference between CLL and CL is also indicated by the amount of data and the number of reports that the contractor is required to submit to the customer during the contract performance period. These requirements are listed as Contract Data Requirements Lists (CDRLs). A typical CL has about 20 CDRLs, whereas a CLL has about 90. The CL CDRLs deal with data essential for the compatibility and integration of the satellite to the launcher. On the other hand, CLL CDRLs require the contractor to submit much more

---

[17]Colonel Thomas W. Meyer, Director of SDIO Directed Energy Office, said that using a commercial procurement for launching the LACE/RME satellite saved an estimated $13–$17 million over the cost of an Air Force (commercial-like) procurement. Since SDIO paid $38 million for the launch, the commercial procurement is on average 28 percent lower than CLL. (Edward H. Kolcum, "SDI Laser Test Satellites Placed in Precise Orbits," *Aviation Week & Space Technology*, February 19, 1990, pp. 24–25.)

[18]The Air Force estimated that the Navy's procurement approach (commercial) would have a total cost of $729.4 million, assuming no launch failure, and the Air Force's approach (MLV-2 or commercial-like) would lead to a total cost of only $609.0 million. The Navy stated that the price that Hughes paid for launch services is $709.4, not $729.4 million. Moreover, the Navy claimed that the Air Force neglected to include $192 million related to insurance, launch and on-orbit testing, etc. The total program using the commercial procurement would have been $709.4 million, whereas a program using a commercial-like procurement would have been $801 million. The UCSLcl,cll is 11 percent. With the Atlas launch reliability record to be about 87 percent, it is more likely to have one failure than none. Based on the Navy data, we estimated that the savings of using commercial procurement could be $135 million, or 16 percent, for ten launches with one expected failure. "UHF Follow-On Launch Cost Studies," transmitted by Donald J. Yockey, Principal Deputy, the Office of Under Secretary of Defense for Acquisition, to Senator Sam Nunn, Chairman of Committee on Armed Services, on April 6, 1990.

detailed launcher technical, cost, and progress data, which might not be useful in monitoring a CL or CLL fixed-price contract. The government wants to know about and approve engineering waivers, deviations, and changes. It asks for a series of reports on monthly cost variance, future funding needs, and even plantwide indirect cost data. With years of experience in launcher manufacturing and processes, the major launch contractors have the capability to decide what alterations are safe. Moreover, the free reflight provision in a CLL contract should provide adequate incentives for the contractor to be very cautious in trading high reliability for lower cost. The fixed-price contract further protects the government's financial interest from contractor's cost overrun. Thus, most of those cost reports seem to serve little purpose. The government requirement of a Contract Work Breakdown Structure (WBS) and associated labor performance and cost data also places a burden on the launch contractor, because it normally manages its program through functional organizations (such as accounting, marketing, production, purchasing, etc.) rather than by WBS. The contractor is forced to keep two tracking systems instead of one. All these requirements play a significant role in making these contractors allocate about half of their labor for preparing and managing paperwork and only half for the actual hands-on production and processing of launchers. Based on our interviews with launch contractors and other participants in the launch community, we believe that a cost savings, especially at the lower end of the 16 percent to 28 percent range, in going from CLL to CL is achievable.

Another indicator is the complexity of the RFP. A commercial RFP can be less than 100 pages, whereas a commercial-like RFP runs over 1000 pages filled with Federal Acquisition Regulations, military standards, and documentation requirements. Not only is the preparation of a bid much more costly for CLL, the commitment made in the bid and resultant contract will inevitably demand high labor resources and cost to fulfill.

VALUES OF FIC

The government investigation cost of launch failure depends on the duration of investigation. We assume that the duration spans the full period of launch standdown. The standdown times are 2.8, 4.1, and

5.2 months for Delta, Atlas/Centaur, and Titan, respectively.[19] We further assume that the Air Force personnel involved in the investigation amount to the equivalent of 100 full-time persons and that the annual salary and other expenses per person are $200,000. Using these figures, we arrive at $5, $7, and $10 million for Delta, Atlas/Centaur, and Titan, respectively.

MODEL APPLICATION TO THE TITAN IV CONTRACT

Our model is first applied to the Titan IV contract, which we classify as a government launch procurement. Would the Air Force be better off if a commercial-like launch procurement were used instead? Unlike the other two major Air Force launch contracts, MLV-1 and -2, Titan IVs are procured to launch satellites with many different missions. Our example here is the launch of Block-IV DSP (Defense Support Program) satellites. Table 4.4 shows that the program requires the launch of five DSP satellites at a rate of one per year. Titan IV reliability is 93.6 percent. Each government Titan IV launch costs $218 million, and the DSP satellite costs $221 million.

As discussed for Titan IV in Chapter Three, the Titan IV contract is well approximated by the no-insurance category in which the contractor is not liable for the loss of launcher or satellite in the event of a launch failure. Since DSPs and other Titan IV payloads tend to be expensive and critical satellites, the Air Force uses the GL contract so that the manufacturing and processing of the launchers are under close monitoring. The Air Force could have considered CLL procurement, where government monitoring is still substantial. This is the case we examine below. We, however, do not consider CL, because, at the time and even now, the Air Force would be highly unlikely to leave these important military payloads fully to the contractor.

As pointed out in the section on Values of Unadjusted Cost Savings per Launch, an important parameter is $UCSL_{cll,gl}$, which is assumed to range from 0 to 20 percent. Figure 4.2 shows that the total cost savings in using CLLs was unlikely to be large. Even in the case most

---

[19]Harry Bernstein, "Space Launch Systems Resiliency," *Proceedings of the Twenty-Fifth Space Congress,* Cocoa Beach, Florida, April 26–29, 1988.

RAND #460-4.2-05/93

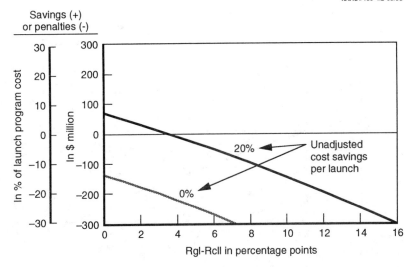

Figure 4.2—Savings or Penalties in Choosing Commercial-Like Launches
(DSP-BL IV on Titan IV and no insurance on launchers and satellites
for CLLs)

optimistic to CLLs, in which the unadjusted cost savings per launch
is 20 percent and the CLL launch reliability is as high as GL's, the
savings amounted to only $70 million or 7 percent of the launch pro-
gram cost. The launch savings shrank from the unadjusted 20 percent to 7 percent, because the cost of keeping a backup launcher/
satellite for CLLs amounted to 13 percent of launch cost. The high
backup cost is a result of having merely five DSPs in the launch program and the infrequent launch of only one DSP per year. The
longer the launch program, the higher the cost of buying the pair of
launcher and satellite early, because of the time value of money.
Moreover, the fewer the number of satellites in the program, the
higher the backup cost per launch, because there are fewer satellites
or launches to spread the backup cost over.

The Air Force could further maintain that, if it selected CLLs, it would
insist that the contractor pays for the loss of launchers and satellites
resulting from launch failures. With an extra insurance fee rate of 4
percent, as discussed in the section on Model Description, the $70
million savings would vanish even in the most optimistic CLL case
(Figure 4.3).

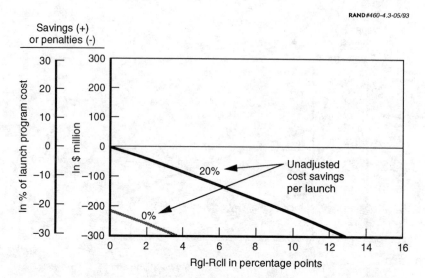

Figure 4.3—Savings or Penalties in Choosing Commercial-Like Launches
(DSPS-BL IV on Titan IV and insurance on launchers and satellites
for CLLs)

We have found that savings in using CLLs for Titan IVs would have
been small or nonexistent. Moreover, at the time that the Titan IV
contract was signed, commercial-like launches had not begun, and
CLLs' reliability record was unknown. Thus, the Air Force had good
reason to have chosen government launches for Titan IVs.

MODEL APPLICATION TO THE MLV-1 CONTRACT

In 1987, the Air Force used a commercial-like launch procurement
for the launches of 20 GPS satellites under MLV-1. Table 4.4 shows
that the launch rate is five per year. The Delta reliability at the time
that the MLV-1 was made was 94.3 percent. Each commercial-like
launch of Delta II costs $59 million, and the GPS satellite costs $46
million. As discussed in the section on MLV-1, the provision cover-
ing launch failures is equivalent to an insurance on launchers, but
not satellites. We now examine whether the Air Force could have
saved money by using commercial, instead of commercial-like,
launch procurement.

In the section on Values of Unadjusted Cost Savings per Launch, we noted that the important parameter, UCSLcl,cll, ranges from 0 to 28 percent. Figure 4.4 shows the total cost savings as a function of unadjusted cost savings per launch and the reliability differential. The figure shows a large triangular region (above the horizontal axis) where a commercial procurement of MLV-1 would have produced considerable savings. At the most optimistic case for CLs, the savings could have been $280 million or 25 percent of the launch program cost. Even with a UCSLcl,cll of 16 percent, the savings could still be $140 million or 13 percent. For MLV-1, the UCSL only has to be adjusted downward by about 3 percent, because the backup cost is spread over a large number of satellites (20).

The Air Force might have insisted the contractor be liable for the loss of satellites as a condition of using CLs instead of CLLs. It could argue that the less its monitoring is, the less it wants to assume the risk of loss. Figure 4.5 shows the case in which the contractor carries insurance on both launchers and satellites in the CL contract. The extra insurance fee on satellites reduces the savings by about 3 percent, but the figure shows that the likelihood of savings for selecting CLs is still rather high. The Air Force could have saved money by using CLs, instead of CLLs.

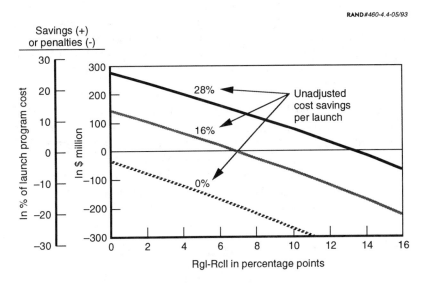

Figure 4.4—Savings or Penalties in Choosing Commercial Launches
(GPS on MLV-1 and insurance on launchers for CLs)

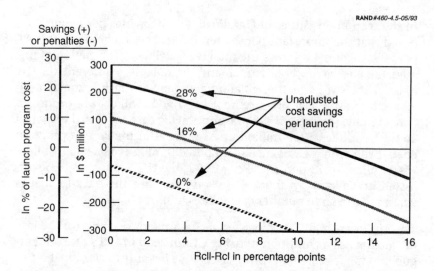

Figure 4.5—Savings or Penalties in Choosing Commercial Launches
(GPS on MLV-1 and insurance on launchers and satellites for CLs)

On the other hand, if the Air Force's assessment is accurate, UCSLcl,cll could be zero or close to it. Then, selecting CLs and imposing backup launcher/satellite and/or satellite insurance would result in a loss in using CLs as opposed to CLLs. We recommend that, in future launch contracts, bidders be requested to give prices on both CLs and CLLs. These prices can then be used to calculate the important parameter UCSL directly, and the Air Force will know whether it is closer to 0 or 28 percent.

In retrospect, since both CL and CLL Delta IIs have had a perfect launch record, the Air Force could have saved money by using CLs, instead of CLLs, for MLV-1. On the other hand, the first commercial launch did not occur until 1989, whereas the MLV-1 contract was signed in 1987. Without any commercial launch experience, it was quite reasonable for the Air Force to select commercial-like, instead of commercial, launches.

MODEL APPLICATION TO THE MLV-2 CONTRACT

In 1988, the Air Force used a CLL procurement for MLV-2. The seven DSCS-III satellite payloads are to be launched over four years. Each

Atlas II launch costs $61 million and each satellite, $84 million. The insurance provisions of MLV-1 and -2 contracts are similar: the contractor will provide free reflight or refund for a launch failure, and the loss of a satellite is not covered. The reliability of Atlas II was 87.2 percent at the time the decision on MLV-2 was made.

Comparing Figures 4.6 and 4.7 for MLV-2 with Figures 4.4 and 4.5 for MLV-1, we found that the triangular region of savings for MLV-2 is much smaller than that for MLV-1. The main reason is that the seven payloads in the MLV-2 launch program are considerably smaller than the 20 payloads in MLV-1. Therefore, MLV-2 has fewer payloads to spread the backup launcher/satellite cost. When insurance is extended to satellites (Figures 4.5 and 4.7), MLV-2 fares worse because DSCS-III satellite ($84 million) costs almost twice as much as a GPS satellite ($46 million). Thus, if the Air Force insisted on having insurance on both launchers and satellites in order to risk trying commercial launches, higher satellite cost and thus a higher extra insurance fee rate would make CLs for MLV-2 less attractive than CLs for MLV-1. In addition to small and even nonexistent savings, again there was no CL launch record when the MLV-2 contract was signed.

RAND#460-4.6-05/93

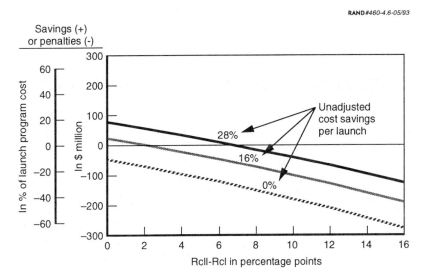

Figure 4.6—Savings or Penalties in Choosing Commercial Launches
(DSCS-III on MLV-2 and insurance on launchers for CLs)

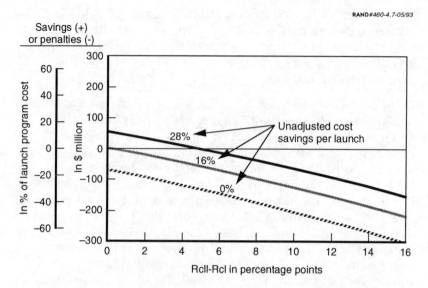

Figure 4.7—Savings or Penalties in Choosing Commercial Launches
(DSCS-III on MLV-2 and insurance on launchers and satellites for CLs)

Therefore, the Air Force again had good reasons to use CLL, instead of CL, procurement for MLV-2.

MODEL APPLICATION TO THE HUGHES/GD CONTRACT

In 1988, the Navy signed a contract for the production and launch of ten UHF-FO satellites. Hughes subsequently signed a contract with General Dynamics for the launches at a rate of two per year. This is a commercial launch contract with insurance on both launchers and satellites. A UHF-FO satellite costs $54 million. If the launch cost and reliability for CLLs are $61 million and 87.2 percent, the launch cost and reliability for CLs could be different. Figure 4.8 reflects the sensitivity of savings or penalties to CLs' launch cost savings and re- liability differentials. By comparing Figures 4.4 and 4.8, we found the likelihood and the size of savings lie between MLV-1 and -2.

The unadjusted cost savings per launch in using commercial launches based on Navy data is 16 percent. At the same time, the Navy considers that commercial launches are just as reliable as commercial-like launches (see section on Navy for historical back-

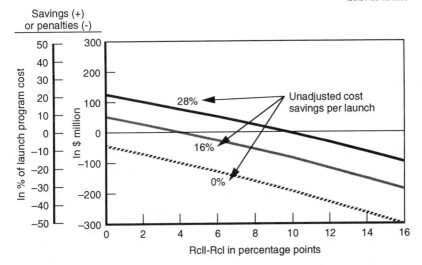

Figure 4.8—Savings or Penalties in Choosing Commercial Launches
(UHF-FO on Hughes/GD and insurance on launchers and satellites
for CLs)

ground on choosing CLs). This leads to a launch savings of 9 percent.
We found in the section on Comparison of Reliability Records of
Three Types of Launches that CLs and CLLs with the same reliability
cannot be ruled out at the 95 percent confidence level. Thus, the
Navy belief in no reliability differential is understandable and leads
to the use of CLs instead of CLLs. In fact, the savings could be higher.
The Navy finds no need for backup launcher or satellite in CLs. If no
adjustment is made to the launch costs, the launch savings would be
the same as the unadjusted savings, or 16 percent.

MODEL APPLICATION TO THE MLV-3 CONTRACT

We have applied the model to four major launch contracts—Titan
IVs, MLV-1, MLV-2, and Hughes/GD—and found that in all four
cases the Air Force and the Navy have good reasons for the launch
procurement types that they have chosen. We now apply the model
to the upcoming MLV-3 procurement for GPS-FO satellites.

The RFP issued on September 16, 1992, stipulated a commercial-like
launch procurement. The Air Force anticipates 20 launches at an average rate of 4 launches per year with a flexibility of two additional

launches per year. We assume a launch reliability of 95.4 percent.[20] The RFP also stipulated a 60-day callup, which requires the contractor to launch a satellite with 60 days of advance notice. The contractor will likely meet the requirement by stockpiling unassembled launch components at the launch site. Our backup arrangement specified for commercial launches will meet this callup provision quite well. Thus, the Air Force wants, in effect, a similar backup arrangement for commercial-like launches also. Here, we assume that there is one additional backup in CLs.[21] If that is not needed, the CL results would be more attractive than those discussed here.

Figures 4.9 and 4.10 show that there is a large region where commercial launches can be more economical than commercial-like launches. For example, in Figure 4.9, if the unadjusted cost savings per launch is 16 percent and if the reliability of CLs is not more than 6 percentage points below that of CLLs, CL procurement for MLV-3 will be less expensive than CLL procurement, as the Air Force currently anticipates. If CLLs and CLs have the same reliability, the savings of CL procurement can be $130 million or 12 percent of the MLV-3 launch program cost. On the other hand, if the Air Force is correct in estimating that the $UCSL_{cl,cll}$ is 1, a CL procurement can be 4 percent more, even when CLs and CLLs have the same reliability.

We suggest that a good way to find out is to request the bidders on MLV-3 to quote prices on both CLs and CLLs. Then, the Air Force will know whether the UCSL is close to 1 or other values. This important piece of information will help the Air Force to determine the potential savings or penalties in choosing CLs.

---

[20]In the section on Model Sensitivity, we will show that the results are insensitive to other reliability values. We will also use MLV-3 to demonstrate that our results and recommendations are not sensitive to the discount rate, which we have assumed to be 10 percent.

[21]We assume, however, that the cost of storage areas for the additional backup at the launch site is insignificant. Spare storage space would probably be available. If not, the construction cost for additional storage is small. The cost for satellite storage would be about $200,000 and that for an unassembled space launch vehicle, $300,000. The total cost would be $500,000. To be conservative, we assume it to be $1 million. If the building cost is amortized over 30 years at a 10 percent discount rate, the annual amortized cost is $100,000. For example, spread over four GPS-FO launches per year, the added cost to each launch is $25,000. Since a GPS-FO launch is expected to be about $60 million, the added storage cost amounts to only 0.04 percent and is negligible.

RAND #460-4.9-0593

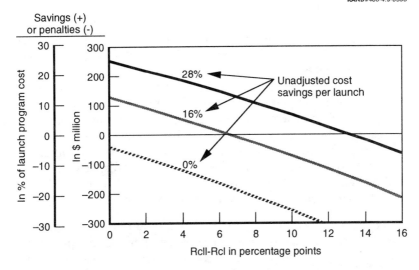

Figure 4.9—Savings or Penalties in Choosing Commercial Launches
(GPS-FO on MLV-3 and insurance on launchers for CLs)

RAND #460-4.10-0593

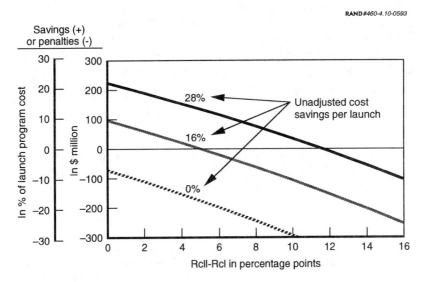

Figure 4.10—Savings or Penalties in Choosing Commercial Launches
(GPS-FO on MLV-3 and insurance on launchers and satellites for CLs)

The pricing data will also allow the Air Force to make a hybrid comparison. The Air Force is worried that contractors other than McDonnell Douglas do not have a long reliability record in the MLV-3 lift range. The Air Force's choice of CLLs would allow adequate oversight regardless which bidder wins. This seems fair, but actually may impose too much oversight if a contractor with a good record wins. There is no need to apply the same level of government oversight to all contractors regardless of their experience and record. We recommend a hybrid approach that compares commercial-launch bids from contractors with good records with commercial-like launch bids from contractors with short records. This way, the Air Force can alleviate its worries in case an inexperienced bidder wins and take advantage of a more experienced contractor.

For the same reason, bidders on MLV-3 should request pricing data for options on insuring launchers and/or satellites. Some parties are concerned that new competitors may be excluded because they cannot obtain insurance. Those who cannot obtain outside insurance can self-insure. If a new competitor can neither obtain outside insurance nor is financially strong enough to make its self-insurance credible, the government will incur a high risk in selecting such a contractor for the launches. The MLV-3 RFP excludes the option of buying insurance on satellites. We believe the exclusion has a higher likelihood of excluding new competitors because the Air Force may not be willing to risk the financial loss of satellites.

Since the RFP has been issued and has ruled out CLs, MLV-3 will not go commercial, unless the Congress or the new Administration intervenes.

In addition to the reliability and cost tradeoff, there are other factors to consider in deciding which launch procurement to use. These factors are discussed in Chapter Five.

MODEL SENSITIVITY

We have used the reliability of Delta launchers as a proxy for the above calculations, since MLV-3 is still under procurement and no contractor has been selected yet. In Figure 4.11, we have made a sensitivity run using a reliability that is as much as 10 percentage points lower or 85.4 percent. It is clear that our results are insensitive to the value used for the CLL reliability for all three cases of

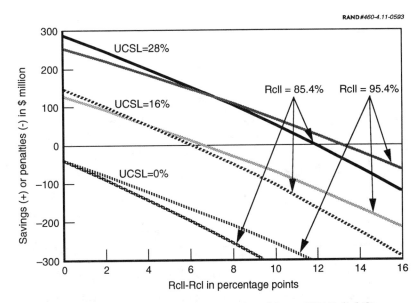

RAND #460-4.11-0593

Figure 4.11—Sensitivity of Savings or Penalties to CLL Reliability
(GPS-FO on MLV-3 and insurance on launchers for CLs)

UCSLcl,cll. The results are sensitive to the relative reliability difference of CLLs and CLs, Rcll – Rcl, not to the absolute value of Rcll.
The insensitivity of results to reliability also applies to the other four
cases in Table 4.3.

Figure 4.12 shows that our results are not sensitive to the discount
rate used. The reason is that, for the five launch programs studied in
this report, a program lasts at most five years. For such a short period, the discounted launch program cost is not very sensitive to the
discount rate. Moreover, the comparison of CLs and CLLs amounts
to the difference of two program costs, and the difference is even less
sensitive to the discount rate.

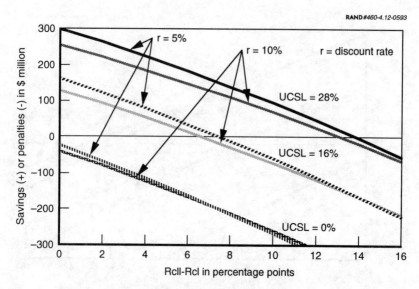

Figure 4.12—Sensitivity of Savings or Penalties to Discount Rate
(GPS-FO on MLV-3 and insurance on launchers for CLs)

Chapter Five

GUIDELINES FOR SELECTING COMMERCIAL LAUNCH PROCUREMENT

In this chapter, we will first discuss a general approach to launch commercialization. We then apply the guidelines to the upcoming MLV-3 launch procurement.

A GENERAL APPROACH TO LAUNCH COMMERCIALIZATION

For a particular satellite program, how should one select among a commercial, commercial-like, or government launch procurement? And, within each procurement type, there are variations in government oversight and contractor compliance. We hope that the guidelines developed here will help us decide the appropriate level of oversight and compliance. Many of the arguments for close government supervision of launch manufacturing and processing are based on the concern that the launch contractors have limited experience, having launched satellites on their own only since August 1989. In reality, launch contractors have been working actively side by side with NASA and Air Force personnel in launch activities since space launches began. There are also security concerns for some highly classified payloads, although the U.S. launch providers have long participated in the processing of these payloads. Some further argue that, even if the launch industry can launch commercial satellites on its own, it needs government supervision to launch certain government payloads, because the launch vehicles might have to be tailored for specific military payloads or because the military payloads are simply too expensive and time critical to be totally entrusted to launch contractors. The implicit assumption is, again, that government supervision improves reliability or that the government cannot afford to find out whether contractors can provide the same reliability on their own. To alleviate these concerns, we propose an approach consisting of the following elements:

- Adopt an evolutionary approach to launch commercialization.

- Emphasize contractor incentives, instead of government oversight, to raise reliability and lower cost.

- Use a set of guidelines for selecting which payloads should use commercial launches.

An Evolutionary Approach

The government has long been interested in off-the-shelf procurement of services and equipment. Space launch services have many attributes that make them particularly suitable for commercial procurement. First, their costs, performance and reliability can be measured easily, so the government can monitor them to determine whether the capability and record continue to satisfy requirements. Second, few payloads, military or otherwise, should require the modification of space launch vehicles. For ground vehicles, airplanes, and ships, we might need extra speed, armor, and durability for military applications. For space launch vehicles, the United States has not been worried, thus far, that its adversary would attack space launch vehicles in flight. The hardness, vibration-tolerance, and other requirements for military and commercial launch vehicles can be the same, although there should be room for exceptions for military or commercial payloads. It would, however, be a mistake to design military payloads that routinely reach the limit of our heaviest-lift launch vehicles and, worse yet, require modification of, or unique processing procedures for, the launch vehicle. With a declining defense budget, the DoD cannot afford many launch vehicles designed exclusively for military use. On the other hand, the short commercial launch record, especially for the Titan IV class, rules out the possibility of total launch commercialization at this time. Therefore, a program leading to more launch commercialization, with a pace commensurate with the evolving commercial launch record, seems reasonable and feasible. We should, however, be open to those exceptions that require close government launch monitoring.

Some might think that the most conservative pace of commercialization for the DoD to follow would be government launches for government payloads, and commercial launches only when the launch record clearly meets the government's reliability standards. There is a problem with this approach. From Table 3.3, launches for commercial customers account for only a third of the total U.S. launches. If half of the government launches were procured commercially, it would double the number of commercial launches. In other words, the time to reach the same number of commercial launches would be half. Thus, without any help from the government, it may take twice as long to establish a commercial launch record. For the Titan IV class, it might take decades to establish a reasonably confident commercial launch record for the Titan III, the Titan IV's commercial counterpart. Moreover, the continued heavy investment by Arianespace on its Ariane 5 and the emerging launch competition from Japan, PRC, and CIS (or its republics) do not allow us to wait much longer. We need government payloads to establish promptly a commercial launch record.

Emphasize Contractor Incentives Instead of Government Oversight

A persistent argument for close government control and monitoring is that certain government payloads are much more expensive or time critical than commercial payloads. Therefore, government involvement is needed to ensure high or higher-than-commercial launch reliability. Unfortunately, government involvement can increase the cost structure not only for government launches but also for commercial launches. The launch providers cannot afford to use two different management systems, manufacturing procedures, and launch processing for government and commercial launches. Since the government accounts for about two-thirds of the launches, the launch providers will use essentially the same system and procedures that met the government requirements for commercial launches. Another example of government oversight that affects commercial launches is test control. If the government wants an added test, who pays for the testing and delay? The government might be willing to pay only for the direct cost of the test and for the added cost to the government launch resulting from the delay. On the other hand, the contractor is worried that an added test demanded by the government will delay not only the government launch at hand but also the commercial launch that follows. The contractor wants the government to pay for the delay penalty that the contractor has promised to the commercial customer. The contractor has to agree to such contract provisions with commercial customers, because not agreeing to them might place U.S. launch providers at a disadvantage. Our chief competitor, Arianespace, has such a provision and processes its launches, commercial or military, essentially through the same set of tests. Instead of oversight, measures that encourage contractor incentives can ensure reliability. The purchase of insurance on launchers and satellites is one such measure.

Although the government traditionally uses self-insurance to save on fees and commissions, the launch procurement trend is toward increased purchase of insurance. In our discussions of MLV-1 and -2, we explained that the Air Force has included insurance on launchers in these contracts. The Navy's UHF-FO program has insurance on both launchers and satellites. Thus, a precedent has been set for allowing the purchase of insurance on launchers and satellites. How does insurance help?

It helps in two ways. First, our interviews indicated that some satellite and launch managers might be more risk-averse than the insurance industry. Such managers would, therefore, insist on government approvals of many checkpoints and on contractor submission of numerous documents, even if excessive monitoring did not cost-effectively enhance reliability. Requiring that a launch contractor purchase insurance could be cheaper. At the same time, the contractor also has a strong incentive to have successful launches. Otherwise, the company's future insurance premiums will be higher, or its insurance policy might even be canceled. Second, if DoD insists that the bidders provide quoted prices that include an insurance premium, the insurance industry will price various launchers with different launch reliability records for DoD and will assume the uncertainties and risks, which many DoD managers are unwilling to take.

Guidelines for Selecting Commercial Launches

The Air Force has selected commercial-like procurement for the launches of GPS and DSCS through MLV-1 and -2, respectively. It has, however, selected government procurement for the launches of a variety of payloads through Titan IV procurement. The Navy and

SDIO used commercial procurement for UHF-FOs and the LACE/RME satellite, respectively. What guidelines should they follow in the future to procure launch services? In what situations would commercial procurement be advisable? We will list the key guidelines in terms of questions. Later, we will apply the guidelines to the upcoming MLV-3 contracts.

- Does the procurement selection model indicate a break-even point so favorable that commercial launches are likely to be less costly? The model shows the tradeoff between potential cost savings and lower launch reliability.

- Is it necessary to modify the launch vehicle to accommodate the payload?

- If there is a delay in delivery to orbit, how seriously would it affect the timeliness and quality of mission performance?

- If commercial launch is selected, is it feasible to switch back to commercial-like or government launch in the event that commercial launches fare poorly? If so, how many launch failures are tolerable before switching back?

Let us elaborate on these guidelines. If the procurement selection model gives an unfavorable break-even point, the benefits of commercial launch may not be large or certain. If the launch vehicle needs to be modified, DoD will worry that the modification would make the past launch reliability less relevant. Therefore, an added amount of government supervision might be required. If the mission is time critical, DoD will be concerned about launch failure or delay. Finally, if DoD finds it impractical to switch back to more oversight in the event that the commercial launch of the initial payload fared poorly, DoD may not want to try commercial launch in the first place.

The issues brought out by these questions should help us to determine the pace of launch commercialization. We will discuss the pace after we apply these questions to the upcoming MLV-3.

SHOULD MLV-3 USE COMMERCIAL PROCUREMENT?

By 1993 or 1994, the launches of GPS Block II/IIA for the full constellation (21 satellites plus three on-orbit spares) will be complete. The

Air Force is planning to procure services for launching replenishment satellites, GPS Block IIR, beginning in fiscal year 1996. We believe that the MLV-3 procurement is vital to the direction and pace of launch commercialization for two reasons. First, there are few existing and planned programs that involve a large number of launches. The MLV-3 contract will cover 20 launches. If the Air Force chose commercial procurement for these launches, it would clearly show its support of the recent DoD and national space policy directives to use commercial services to the maximum extent possible. Second, unlike the United States' largest launch program, Titan IV, the launchers for GPS are within the most relevant lift-class range for commercial launchers. Therefore, if MLV-3 were to be procured commercially, it would have the strongest influence on the U.S. launch industry's competitiveness. It would mean a large added demand for commercial-class launchers without the added cost of compliance.

The current indication is, however, that the MLV-3 launches will be procured in a commercial-like manner as with MLV-1 and -2. Then, the only Air Force launch programs that can be treated as commercially procured are those associated with small launch vehicles. Unfortunately, these small launchers do not have the capability to deliver geosynchronous communication satellites, which are the most important ones for the commercial markets. Without strong support from the Air Force, launch commercialization will be slow and might even wilt. On the other hand, we need to examine what risk the Air Force would incur by using commercial procurement for MLV-3. Let us answer the questions in the previous section in turn.

Procurement Selection Model Results

In the previous chapter, we showed that the likelihood for CL procurement of MLV-3 to be less expensive than CLL procurement is good, especially if CLs have as good a reliability record as CLLs. Potential launchers for MLV-3 include Delta II, Atlas, and, if upgraded, Titan II. Table 3.1 shows that, as of July 7, 1992, Delta II had 11 commercial launches and 15 commercial-like launches; all were successful. Atlas had five commercial launches with one failure. Although not broken out in Table 3.1, Titan II had two government launches; both were successful. Thus, the commercial launch record came essentially from Delta IIs, whose launch record has been excellent. There is no indication that commercial practices have degraded

Delta's launch reliability. Atlas and Titan II have had too few or no commercial launches to provide a commercial reliability record.

Need to Modify Launchers

There could be a modest growth in spacecraft weight that requires a small upgrade of launcher delivery capability. There should, however, be no need to modify Delta, Atlas, or Titan in any substantial way to accommodate GPS IIR.

Impact of Delay on Timely Performance

What is the likelihood for mission delay, if CLs are used for MLV-3? GPS-FOs constitute a unique and attractive satellite system. When launches of GPS IIR satellites begin in FY 1996, the full constellation of GPS II/IIA (21 satellites plus three spares) will have been established for only two years (Table 5.1). Even if commercial launches turn out to be less reliable than commercial-like launches, the delay caused by using commercial launches might not be significant for three reasons.

First, historically U.S. satellites have demonstrated a longer on-orbit life than originally designed. The old GPS IIs, as well as the on-orbit spares, can serve as substitutes if launch delays of GPS IIRs are encountered. Current on-orbit GPS satellites consist of Blocks I and II.

Table 5.1

**Mission Delay and Contract Modification
(GPS-FO under MLV-III)**

- Is mission delay bearable?
  - A GPS constellation (21 satellites plus 3 spares) will have been established
  - Constellation degrades gracefully despite launch failures
    - Two-dimensional position fixing still feasible
    - Three-dimensional many hours a day
- No substantial contract modifications

Block I satellites are already lasting on average three years longer than their designed mission duration of four years.[1] Since some of them are still operating, the actual average mission duration will further exceed the design life. Block II satellites are still too young to know their actual average life, but they carry ten years of consumables, indicating that the designer expects them to last beyond the six-year mean mission life. Some could argue that old GPS satellites are not perfect substitutes for the GPS-FOs. For example, nuclear hardening is being added to the GPS-FOs, to be more capable in a hostile wartime environment. The collapse of the Soviet empire makes a conflict involving nuclear effects on satellites highly unlikely. It is even more unlikely that the conflict would occur at a time when launch delays of GPS-FOs are encountered and the old satellites are being used as substitutes. Even if the satellites should be hardened against nuclear effects, one should ask whether the degree of hardening implemented in GPS-FOs would be adequate against nuclear interference and whether the former Soviet republics could destroy hardened GPSs. Similarly, although auto-navigation for six months is a desirable feature, GPS ground stations in the United States are unlikely to be under attack. Even if they were, a few stations for monitoring and uplinks might still be available for maintaining the few old GPS satellites being used as substitutes.

Second, even if GPS IIs were not available, the position-fixing capability would degrade gracefully. Lacking a full constellation, the GPS satellites can still provide a two-dimensional (2D), as opposed to a three-dimensional (3D), position fixing. Also, 3D is feasible for many hours of the day, as opposed to almost 24 hours a day. For many applications, 2D will suffice. For example, since the altitude of ships is zero, 2D position fixing of ship location is adequate. So is location finding of ground soldiers. For aircraft, the altitude can be determined by other means such as radar. Instead of 18 satellites for continuous 3D, 15 satellites are enough to perform continuous 2D. In fact, even with as few as 12 satellites, they can conduct 20 hours of 2D fixing and 12 hours of 3D.

Third, our suggestion of adding a backup or standby payload/launcher to the commercial launch procurement would further re-

---

[1]Thomas S. Logsdon, Global Positioning Systems: Principles and Applications, an extension course presented at University of California, Irvine, October 1–4, 1991.

duce the chance of delay. It should be emphasized that our cost comparison of commercial and commercial-like launches has included the standby cost for a commercial launch but not for a commercial-like launch. If a program manager does not find a standby for a commercial launch necessary, then the commercial launch could be even less expensive and more attractive than a commercial-like launch.

Penalties of Switching from CLs to CLLs

A general contention is that commercial launches are less reliable because they have less government oversight. The commercial launch record is too short to confirm or deny the assertion with high confidence. To address such contentions, we can add a switching or cancellation provision in a commercial launch contract. If the Air Force decided to use commercial procurement for MLV-3, but commercial launches fare far worse than expected, the Air Force would have the right to reimpose the types of government oversight that are common in a commercial-like procurement for the remaining launches or to cancel the contract. How practical is a midcourse switch from commercial launches to commercial-like launches? How large would be the penalty for the switch or cancellation?

Since the current plan is to procure MLV-3 in a commercial-like manner, it will take considerable effort to convince the Air Force to use commercial procurement instead. If the Air Force finally procured commercially, it would be undesirable to switch out of commercial procurement prematurely, because the Air Force might not try commercial procurement again for future launch programs in this commercially important medium-lift class. To give commercial procurement the best chance to succeed, we believe that a switch should not be made until the Air Force has high confidence that the commercial launch reliability is worse than expected. Table 5.2 illustrates the decision rule for when to switch. In the illustration, we assume that the launch program manager would switch when there is a 5 percent or less chance that the commercial reliability is as good as or better than originally expected. In the table, we have made an adjustment to the decision rule, namely, that one will never switch

Table 5.2

Decision Rule for Quitting Commercial Launches

(Quit if there is only 5 percent or less chance that the commercial reliability is as high as or higher than expected. Do not quit, however, if number of failures is two or less.)

Number of Launches (N)	Quit if Number of Failures Is Equal to or Exceeds					
	R = 95%*	R = 90%	R = 85%	R = 80%	R = 75%	R = 70%
1	Won't quit** (5%)***	Won't quit (10%)	Won't quit (15%)	Won't quit (20%)	Won't quit (25%)	Won't quit (30%)
2	Won't quit (.25%)	Won't quit (1%)	Won't quit (2.25%)	Won't quit (4%)	Won't quit (6.25%)	Won't quit (9%)
3	3 (0.01%)	3 (.1%)	3 (.34%)	3 (.8%)	3 (1.6%)	3 (2.7%)
4	3 (.05%)	3 (.37%)	3 (1.2%)	3 (2.72%)	4 (.39%)	4 (.81%)
5	3 (.12%)	3 (.86%)	3 (2.66%)	4 (.67%)	4 (1.56%)	4 (4.08%)
6	3 (.22%)	3 (1.58%)	3 (4.73%)	4 (1.70%)	4 (3.76%)	5 (1.09%)
7	3 (.38%)	3 (2.57%)	4 (1.21%)	4 (3.33%)	5 (1.29%)	5 (2.88%)
8	3 (.58%)	3 (3.81%)	4 (2.14%)	5 (1.04%)	5 (2.73%)	6 (1.13%)
9	3 (.84%)	4 (.83%)	4 (3.39%)	5 (1.96%)	5 (4.89%)	6 (2.53%)
10	3 (1.15%)	4 (1.28%)	4 (5.00%)	5 (3.28%)	6 (1.97%)	6 (4.73%)
11	3 (1.52%)	4 (1.85%)	5 (1.59%)	6 (1.17%)	6 (3.43%)	7 (2.16%)

Table 5.2—continued

Number of Launches (N)		Quit if Number of Failures Is Equal to or Exceeds				
	R = 95%*	R = 90%	R = 85%	R = 80%	R = 75%	R = 70%
12	3 (1.96%)	4 (2.56%)	5 (2.39%)	6 (1.94%)	7 (1.43%)	7 (3.86%)
13	3 (2.45%)	4 (3.42%)	5 (3.42%)	6 (3.00%)	7 (2.43%)	8 (1.82%)
14	3 (3.01%)	4 (4.41%)	5 (4.67%)	6 (4.39%)	7 (3.83%)	8 (3.15%)
15	3 (3.62%)	5 (1.27%)	6 (1.68%)	7 (1.81%)	8 (1.73%)	8 (5.00%)
16	3 (4.29%)	5 (1.70%)	6 (2.35%)	7 (2.67%)	8 (2.71%)	9 (2.57%)
17	4 (.88%)	5 (2.21%)	6 (3.19%)	7 (3.77%)	8 (4.02%)	9 (4.03%)
18	4 (1.09%)	5 (2.82%)	6 (4.19%)	8 (1.63%)	9 (1.93%)	10 (2.10%)
19	4 (1.32%)	5 (3.52%)	7 (1.63%)	8 (2.33%)	9 (2.87%)	10 (3.26%)
20	4 (1.59%)	5 (4.32%)	7 (2.19%)	8 (3.21%)	9 (4.09%)	10 (4.80%)

NOTES:

*   R is the expected reliability of commercial launches.

**   "Won't quit" means "won't quit even if all launches up to that point failed." The probability in parentheses corresponds to the case where all launches failed.

*** The percentage in parentheses is the probability that the number of launch failures is equal to or exceeds the indicated number, when the commercial reliability is at the expected value, R.

when the number of failures is two or less.[2] Therefore, one will not switch if both the first and second launches fail. In the table, the values of R in the top row show the commercial reliability that the launch manager expected. The first entry shows the number of failures (out of N launches) at or above which the manager will switch. The second entry (in parentheses) shows the probability that the number of launch failures is equal to or exceeds the indicated number (the first entry) when the commercial reliability is at the expected value, R. For example, the expected commercial reliability is 90 percent. If there are three or more failures out of eight launches, the manager will switch because the probability for the commercial reliability to be 90 percent is only 3.8 percent or less.

It is reasonable for a launch manager to expect the commercial launch reliability to be 90 percent or better. The current commercial launch record, although it is short, has such reliabilities. Here, we assume the expected commercial reliability to be 90 percent.[3] The manager might give up on commercial launches when there are three launch failures in eight or fewer launches, four failures in 9 to 14 launches, or five failures in 15 to 20 launches. What are the losses in such a switching strategy?

The monetary loss and mission delay will depend on when the Air Force gives up on CLs. The launch schedule as specified in the MLV-3 RFP allows two additional launches per year. Thus, if there are one or two launch failures in a year, the launch flexibility in the launch contract can make up for the time delay resulting from failure investigation and corrective actions. A very bad situation is three failures in a row just before switching from CL, when the likelihood of CL reliability to be at or above 90 percent falls below 5 percent. That will cause about a year of delay for failure investigation.[4] In addition, it

---

[2]During the initial few launches, even two failures could indicate that the reliability is unlikely to be high. For example, say one was expecting the commercial launch reliability to be 90 percent. If both of the first two launches failed, the probability for the commercial reliability to be 90 percent or higher would be 1 percent or lower. The 5 percent decision rule would call for a switch. We feel, however, that failures during the first few launches could be caused by the developmental nature of the initial launches. The causes could be quickly corrected and nonrecurring.

[3]Table 5.2 gives numbers for other expected commercial reliabilities. If a launch manager expected the commercial reliability to be as high as 95 percent, he might switch at a lower number of failures.

[4]We do not assume flying through standdown. The length of delay is determined by the standdown time for failure investigation and corrective actions—on average,

may take another year to switch back to CLLs, for a two year delay total. If the switch occurs early in the launch program, some time can be made up after switching by launching two additional satellites a year. If the switch occurs late, the delay will be close to two years.[5] Figure 5.1 shows that the monetary loss ranges from $100 to $180 million if the Air Force purchases insurance only on launchers. The earlier the switch occurs, the smaller the loss. This is no surprise because, if one has to switch, it is better to know it early before many launch failures have occurred. The worst case is a loss of $180 million and a mission delay of two years.

Switching is not an easy task. The Air Force might have to terminate the contract and reprocure the launch services, if it cannot come to

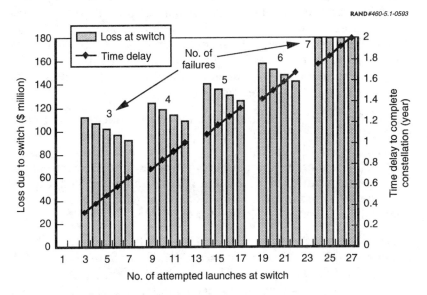

RAND #460-5.1-0593

Figure 5.1—Penalty for Switching from CL to CLL
(GPS-FO on MLV-3 and insurance on launchers)

---

about three to four months per failure. Delta has had 12 failures since 1960, and the average standdown time is 2.8 months. Atlas/Centaur had ten failures during 1962–1990, and the average standdown time was 4.1 months. Harry Bernstein, "Space Launch Systems Resiliency," *Proceedings of the Twenty-Fifth Space Congress*, Cocoa Beach, Florida, April 26–29, 1988.

[5]An alternative to switching is to take the risk and finish the launch program with CLs. This is a viable option when only a few launches are left.

an equitable agreement on price and penalties with the existing contractor. We have made the switch relatively easier by assuming that the Air Force launch program personnel will be maintained.[6] Based on the above discussion, we found that switching under MLV-3 is at least more practical than under other launch programs where the number of launches is much smaller than 20.

A loss of up to $180 million is a sizable sum, but tolerable. Since this cost can be considered the cost of attempting launch commercialization, the government should reimburse the Air Force for a portion, if not all, of the loss. On the other hand, if the commercial reliability turns out to be as good as the commercial-like reliability, the Air Force can save $130 million if the UCSL is 16 percent and $250 million if the UCSL is 28 percent. An option for the Air Force is to purchase insurance on satellites. The extra insurance fee would be $50 million, which would change the above potential savings to $80 and $200 million, respectively. On the other hand, if CLs fare as badly as discussed above, the loss of $180 million will be reimbursed by the insurance company or the launch contractor. The mission delay of up to two years will, however, remain.

Our analysis of MLV-3 indicates that it is a very attractive and important candidate for commercial procurement. If it is procured commercially, it would give U.S. launch commercialization a strong boost. If, however, it is procured commercially and the commercial launches turn out to be much less reliable than expected, the Air Force could reimpose government oversight without causing large financial losses and long mission delays.

We recommend, therefore, that the MLV-3 RFP contain CL and CLL options as well as different contractor liability options, because the Air Force will need the pricing data to ascertain whether CL can have cost savings and whether the financial impact of launch failures can be limited.

PACE OF LAUNCH COMMERCIALIZATION

At one extreme, there have been proposals, such as the Packard Bill, that supported immediate full launch commercialization, with na-

---

[6]We did not reduce the government cost for CLs relative to CLLs in our calculations.

tional security exceptions to be certified by the Secretary of Defense. At the other extreme, there are beliefs that even the current level of commercialization is too much for DoD, which should revert to more control for launcher development, manufacturing, and processing. If the guidelines discussed in this report are followed, MLV-3 might be procured commercially and so should small launchers in general.

On the other hand, our guidelines do not indicate that Titan IV launches are ready for commercialization yet. Let us elaborate on how our four guidelines or questions apply to Titan IVs. It is relatively costly to keep a launcher/satellite backup, because the costs of a Titan IV and its payload are high and there are only a small number of satellites for each mission to spread the backup cost among. Our break-even analysis reflecting these concerns also shows commercial procurement for Titan IVs to be unattractive (Figures 4.2 and 4.3). There have been only three Titan III (Titan IV's commercial counterpart) launches, with one failure. For many classified payloads launched by Titan IV, maximum satellite performance is a major consideration. It could be necessary to modify a launch vehicle or to reduce its safety margin for best accommodation and capacity. A modification of the launcher and a stretching of the safety margin might make the existing reliability record a poor predictor. A launch delay could also significantly affect mission performance. For example, in a time of crisis or war, a few months' delay for adequate surveillance coverage from space could be highly detrimental. Finally, the small number of satellites in each program and their high cost would make the option of switching back to more government oversight procedures impractical and costly.

Figure 5.2 shows that the prices of U.S. launchers are approximately proportional to the launch capacities to geotransfer orbits.[7] Ariane 4 falls on the same line. Unfortunately, there is a problem that prevents Titan III from being competitive with Ariane. Although Titan III can launch two geosynchronous communications satellites at a time, the difficulties of matching customers' payloads and launch schedules have led Martin Marietta to adopt a policy dedicating one Titan III launch to each customer. It no longer matches payloads to

---

[7]The data for Figure 5.2 are from Table 2.1.

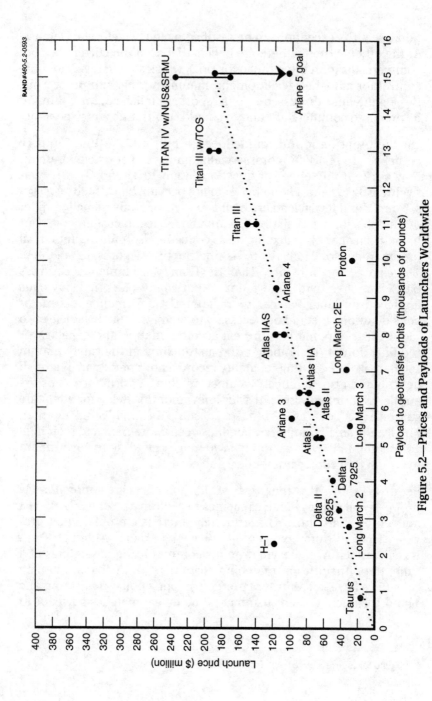

Figure 5.2—Prices and Payloads of Launchers Worldwide

the same launcher for two customers.[8] This policy, in essence, positions Titan III launchers for the much smaller market segment of unusually large payloads. Ariane 5, beginning service by the mid-1990s, is expected to include dual-payload launches and to have a launch cost per pound that is 45 percent lower than the already highly competitive Ariane 4.[9] Without government support for cost reduction and performance improvement, Titan IIIs/IVs may not be competitive in this lift class.

In theory, the United States does not have to be competitive in every lift class. In practice, the United States might have to be in the Ariane 5 class. That Ariane 5 is planned for dual launches has placed it in direct competition with launchers, such as Atlas and future upgraded Deltas, that have about half of Ariane 5's lift capability. Dual launches with Ariane 5s could be less expensive than single launches with Atlas or Delta on a per-payload basis. There is every indication that Arianespace can capture enough launch business to execute its dual-launch policy. Moreover, the new technologies and heavy investments ($5 billion) will give Ariane 5s a low recurring cost. Arianespace has indicated that its goal is to have Ariane 5's price per pound of payload 45 percent lower than that of the current Ariane 4s, which are already quite competitive against the U.S. launchers. There will always be a sizable market for single launches because of more launch scheduling flexibility and less chance of launch delays. If the United States wants to offer a full product line for typical commercial payloads, it needs to develop an Ariane 5-class vehicle to offer dual launches as well. There will also be competition from the Soviet Protons and Zenits, which are in the Ariane 5 and Titan III class. We believe that Titan IVs are not ready for commercial procurement at this time, but that, in the longer term, the United States needs to have launchers competitive with Ariane 5s. These new launchers might be provided by taking advantage of the National Launch System (NLS), currently under development and thus in the

---

[8]It has been reported that Martin Marietta "has restricted its customer base since the first Titan 3 launch by ruling out the pairing of satellites." Daniel J. Marcus, "Martin Marietta Ponders Reviewing Commercial Titan," *Space News,* July 15–28, 1991, p. 20.

[9]At one time, Japan considered developing H-IID, an upgraded version of H-II, to lift between 45,000 and 60,000 lb to LEOs. Ariane 5 will have a capacity of 42,000 lb. "Move over Ariane 5 . . . here comes the H-IID," *Space Business News,* August 20, 1990, p. 1.

process of setting priorities and requirements.[10] We will be more specific in the next chapter.

---

[10]Estimates of launch cost reduction from the NLS or its predecessor, the Advanced Launch Development Program (ALDP), have varied widely over the years. The launch cost per pound has been estimated to be from one-tenth to one-half of the current cost. See, for example, Congressional Budget Office, *Encouraging Private Investment in Space Activities,* February 1991, p. 38.

Chapter Six

STEPS TOWARD STRENGTHENING THE COMMERCIAL LAUNCH INDUSTRY

In previous chapters, we have examined the launch procurement issue. In this chapter, we discuss measures that DoD can take to help strengthen the commercial launch industry regardless of how DoD's launch contracts are procured. DoD should be interested in strengthening the commercial launch industry for three reasons. First, DoD and national space directives already instruct DoD to do so.[1] Second, a launch industry with a strong commercial component will probably charge a more competitive price to DoD launch users and, therefore, will likely lower the launch cost to the government in the long run. Third, a competitive launch industry will help the United States carry out its foreign policies, including fair international trade and missile nonproliferation. Without a commercial launch industry, it would be hard for the United States to argue against foreign subsidies while its launch industry is solely dependent on the government for survival. Also, U.S. efforts to persuade other launch providers not to export launch technologies and components might be construed as a means of protecting the noncompetitive U.S. launch industry.

---

[1]The 1991 National Space Launch Strategy developed by the National Space Council and approved by the President stated that "United States space launch infrastructure, including launch vehicles and supporting facilities, should . . . encourage, to the maximum extent feasible, the development and growth of U.S. private sector space transportation capabilities which can compete internationally." The Vice President's Office, July 24, 1991.

DoD VERY-HEAVY-LIFT DEMAND

In order not to create confusion with heavy-lift vehicles, such as Titan IV, we introduce a very-heavy-lift launch vehicle (VHLLV) class, which is defined as having a lift capability to low earth orbits (LEOs) of 50,000 lb or more.[2] We want to study DoD's needs for VHLLVs, as opposed to the heavy-lift launch vehicles (HLLVs), currently served by Titan IVs, and the medium-lift launch vehicles (MLLVs), served by Delta II7925s and Atlas IIs. This classification, however, does not rule out the possibility that a single family of launch vehicles can serve multiple lift classes. In fact, that was the goal of the National Launch System (NLS).[3] To determine whether a single family can serve the needs of both the Air Force and NASA, we will have to understand their requirements first.

The most significant change in DoD launch demand in the last several years has been in the VHLLV class. Early Strategic Defense Initiative (SDI) architectures contained very heavy spacecraft carrying neutral particle beams and lasers or relay mirrors for ground-based lasers. The shift in missile threats from the former Soviet Union to threats from hostile developing nations has not decreased the likelihood of deploying space-based ballistic missile defenses. The Brilliant Pebbles technologies and their adoption by the Strategic Defense Initiative Organization (SDIO), however, greatly lower the projected DoD demand for VHLLVs. After eliminating these heavy SDI platforms, we found that most DoD payloads projected for the next 20 or 30 years can be delivered by Titan IVs with SRMUs. Thus, if the SRMU program can be successfully completed,[4] DoD's demand for VHLLVs could be very low. Here lies the key difference between DoD's and NASA's launch requirements. NASA

---

[2]Currently, the largest expendable launch vehicle, Titan IV without the Solid Rocket Motor Upgrade (SRMU), can lift 39,000 lb to LEOs, while one with the SRMU could lift 49,000 lb. In comparison, the shuttle has a lift capacity of 51,000 lb.

[3]This study was conducted before the NLS program was canceled.

[4]The design problems and explosions of the SRMU have caused some legislators to propose cancellation of the program. Martin Faga, former Air Force Assistant Secretary for Space, said the composite-cased SRMU was needed to launch four DoD and two NASA payloads. He also said that all four DoD payloads could be launched on Titan IVs with the current steel-cased motors, if one was willing to make cuts in capability and some redesign of the spacecraft. "Exon grills Air Force on launch plans," *Military Space*, May 6, 1991, p. 3.

believes that VHLLVs can lower the cost of launching many payloads for its space station and Space Exploration Initiative (SEI) and other programs. NASA requires a vehicle that can lift 150,000 lb or more to low earth orbits. In fact, General Tom Stafford's Synthesis Group, an intra-agency panel charged with assessing alternate SEI architectures, called for a launch vehicle with a minimum capacity of 330,000 lb and a designed growth to 550,000 lb.[5] On the other hand, the Air Force said it needs a VHLLV capable of delivering 50,000 lb to LEOs by the year 2000—essentially as a replacement for Titan IV.[6] This large difference in the lift requirements between NASA and the Air Force calls for serious compromises in optimization of the engine and other designs for the joint effort to develop a single family of launch vehicles. Not long ago, NASA favored a shuttle-derived vehicle (SDV), which would have a lower development cost ($4 to $10 billion) and be available sooner (7 to 10 years).[7] The Synthesis Group went a step further. It recommended using the old F-1 engines from the Saturn 5 on the first stage. Saturn 5 was the giant booster used in the successful Apollo expeditions to the moon during 1967–1972. On the other hand, the Air Force aims for a new-technology, new-design vehicle that would have a much higher potential for launch cost reduction and responsiveness improvement, in spite of a higher development cost ($10–$15 billion) and a longer development period (10–15 years).

With a tight defense budget for the foreseeable future, a single family of launchers may be desirable. Can it, however, satisfy both NASA and DoD? If so, which launcher design should the United States pursue? The most important element of what NASA and DoD have agreed on thus far is their selection of the Space Transportation Main Engine (STME) for joint development under the NLS program.[8] Still, NASA remains committed to further development of the existing shuttle main engine as a design option for several years, in case the

---

[5]"New launcher can't blast past Beltway," *Space Business News*, June 24, 1991, p. 10.

[6]"Struggles to draw heavy lifter continue," *Space Business News*, February 4, 1991, p. 7.

[7]D. Isbell and A. Lawler, "NASA, U.S. Air Force Contemplate Merger of Launcher Concepts," *Space News*, December 17–23, 1990, p. 3.

[8]J.R. Thompson, NASA Deputy Administrator, said that whereas the shuttle main engine costs $40 million a copy, an expendable STME must cost on the order of $5 million to $10 million a copy. "Struggles to draw heavy lifter continue," *Space Business News*, February 4, 1991, p. 8.

STME encounters development problems.[9] NASA is also continuing work on the Saturn engine.[10] Thus, although DoD and NASA are working hard to make their joint development of an NLS a success, they also recognize the possibility that a single family of launch vehicles might not meet their different needs.

While the National Space Council remains committed to supporting the development of a new generation of launch vehicles, it decided not to select a specific plan for the NLS until 1993. Mark Albrecht, the council's former executive secretary, has said that there are several goals for the NLS, among them "contributing to the long term competitiveness of the commercial launch industry." Our observations now follow.

First, since commercial competitiveness is a key goal, the technologies and innovations useful for VHLLVs and HLLVs should also be readily applicable to reducing the cost of launch vehicles in the most commercially relevant range—from 10,000 to 50,000 lb into LEOs or 2000 to 10,000 lb into geosynchronous orbits (GSOs).[11]

Second, there are some likely vehicle design choices. A modular family of launch vehicles based on the Advanced Launch Development Program (ALDP) is one choice. The number of engines and stages can be altered to lift 10,000 to 500,000 lb payloads into LEOs. Another choice would be a modular family that could lift a much narrower range of 10,000 to 50,000 lb into orbit. It would cover most of the needs for commercial and DoD users, as well as NASA's lighter-lift needs. NASA's very-heavy-lift needs could be served by upgrades of old engines and vehicles, such as those recommended by the Synthesis Group. As discussed above, the cost and time of upgrading could be much lower and shorter than those of developing a new engine and vehicle. The third choice is development of the shuttle-derived vehicle family. It is, however, difficult to see how the SDV family could play a role, because a parallel program would have

---

[9]Douglas Isbell, "NASA-Air Force Panel Settles on ALDP Design, Debates Management," *Space News*, February 18–24, 1991, p. 21.

[10]"National Launch System Garners Support," *Military Space*, December 16, 1991, p. 6.

[11]See Table 2.1 for the LEO delivery capabilities of various current and planned launch vehicles.

to be conducted to develop a low-cost commercial launch vehicle family and, perhaps, some quicker-response military launch vehicles.[12] It is unlikely that the United States can afford two development programs: $4–$10 billion for the SDV family and perhaps another $10 billion for a low-cost, quicker-response launch vehicle.[13]

Third, the United States should consider the desirability and feasibility of international cooperation in VHLLV development. There could be significant cost and technology sharing. VHLLV is particularly suitable for an international joint venture for several reasons.

- VHLLVs will mainly serve scientific space exploration, as opposed to commercial or military purposes. Since there will be so few commercial payloads for VHLLVs, we need not worry how a joint venture would affect future U.S. competition in VHLLVs. We may, however, be concerned about those VHLLV technology flows that would benefit foreign launch providers' lighter, commercial launchers. One way to reduce such flows is for the United States, without foreign participation, to develop a new family of launch vehicles optimized for the most commercially relevant range, 10,000 to 50,000 lb. As discussed above, vehicles in the same range can also handle most U.S. military payloads. The United States will then invite foreign partners for the development of another family of launch vehicles optimized for a range above 50,000 lb. This way, larger engines and other components optimized for VHLLVs might not be economically efficient for lighter launchers. Moreover, this arrangement resolves our major concern that there might be no new launch vehicles in the 10,000 to 30,000 lb range. This concern will be discussed in the next section.

---

[12]In his evaluation of launch vehicle combinations to meet U.S. space traffic needs between 1990 and 2010, Scott Pace found that combinations, including Shuttle-C (a shuttle-derived unmanned vehicle, C for cargo), tend to be more expensive than combinations without Shuttle-C. *U.S. Access to Space: Launch Vehicle Choices for 1990–2010*, RAND, R-3820-AF, March 1990.

[13]An SDV family could serve a role in a situation that requires (a) that the United States decide to seek foreign partners for joint development of a family of VHLLVs, (b) that such partners were found, and (c) that shuttle-derived technologies were chosen for the basic design. We will discuss the feasibility of international development in our next observation.

- A few future U.S. military payloads might have to be launched by VHLLVs, but a joint development does not preclude the United States possessing some of those VHLLVs.

- It would indeed be more conservative, even with the demise of the Soviet Union, to have the joint venture among Western nations. One should, however, consider the inclusion of the CIS. They already have the Energia, which can deliver 220,000 lb into LEOs. Thus, the United States does not have to be concerned that a joint venture will open the door for the CIS to obtain VHLLVs. In the case of VHLLV development, cost sharing might outweigh the risk of transferring technologies to the former Soviet Union.

IMPROVING EXISTING LAUNCHERS AND THE LAUNCH INFRASTRUCTURE

There is a consensus among launch planners that the National Launch System program should include the improvement (but not necessarily the replacement) of existing and planned launchers in what we have described as the most commercially relevant (MCR) range—10,000 to 50,000 lb into LEOs or 2000 to 10,000 lb into GSOs. The differences among NLS planners are in the priorities assigned to the developmental activities most relevant to launchers in this range. Since the NLS development plan is still evolving, it is timely to examine how high such a priority should be. First, it is clear that the commercial launch industry will benefit much more from the improvement of existing launch vehicles and the development of a new family of launch vehicle in the MCR range than those in the VHLLVs (above 50,000 lb to LEOs). Second, even DoD is more likely to benefit economically from lower cost launchers in the MCR range than from the availability of VHLLVs, because most DoD payloads will continue to be in the MCR range. This is particularly true if the historic trend of increasing the performance and weight of military satellites is reversing. At the least, the trend seems to be flattening, if not plainly reversed.

The Air Force has envisioned that the technologies and processes developed under the NLS program can be applied to existing launchers, thus reducing their launch costs by almost one-third. Full cost reduction by an order of magnitude would, however, require redesign of the basic manufacturing and processing of the launch vehicles and their infrastructure.[14]   Moreover, many NLS technologies can be applied to existing launchers by the mid-1990s, whereas a new family of launch vehicles in the 30,000–100,000 lb range (i.e., Titan IV class and above or HLLVs and VHLLVs) can be introduced with an Initial Operational Capability in ten years or so. Although Colonel Roger T. Colgrove of the U.S. Air Force Space Systems Division also mentioned the "option to develop a medium class derivative," there seems to be less consensus among Air Force and NASA planners to develop a new or extended family of launch vehicles in the 10,000–30,000 lb range because it is harder to optimize an engine that spans the wider range from 10,000 lb to 100,000 lb.[15]

We found that the cost savings estimated by the launch industry are consistent with those of the Air Force.[16]   COMSTAC estimated that new component technologies and preplanned product improvements (P3I) can reduce launch cost in existing launchers by over 25 percent. These savings will come from improvements in propulsion and fluid systems, avionics, advanced structures, production, and launch operations. Moreover, these savings can be realized in existing launchers by 1998 and will allow the U.S. launch industry to compete effectively until 2005 or perhaps even 2010. After that, a new family of vehicles will be needed to compete with foreign launch providers. Our recommendations follow.

First, with Ariane 5 coming into service by the mid-1990s and aiming for a cost per pound 45 percent lower than the already highly competitive Ariane 4, we consider that the aggressive and optimistic

---

[14]Colonel Roger T. Colgrove, "Space Launch Roadmap: A Space Systems Division Perspective," November 1990 (quoted in *Space News*, October 29–November 4, 1990, p. 1).

[15]At one time, senior Air Force space officials endorsed the development of a family of new vehicles in the 30,000 to 120,000 lb lift range, but were not willing to raid the $15 billion needed from other programs, such as the NLS. Vincent Kiernan, "Air Force Seeks Outside Support for New Boosters," *Space News*, October 29–November 4, 1990, p. 1.

[16]COMSTAC Innovation & Technology Working Group, *FY 1990 Final Report*, a report to the Commercial Space Transportation Advisory Committee (COMSTAC), October 18, 1990, pp. 13–47.

schedule envisioned by the Air Force and the U.S. launch industry will have to be met.

Second, our greatest concern is that the new family of launch vehicles will only launch payloads down to 30,000 lb into LEOs. Yet the range of 10,000 to 30,000 lb represents some of the most commercially relevant payloads—2000 to 6000 lb for the most important geosynchronous orbits. Therefore, the NLS of similar programs should include the development of new vehicles and the improvement of existing vehicles in this range.

Third, the largest cost component of the NLS program is the development of a new engine. Although a new engine is the key to the new launcher family for both NASA and the Air Force, its applications for improving existing vehicles are relatively much fewer. Fortunately, the funding for technologies and processes most beneficial to existing vehicles would be substantially less than for the new engine. It would be highly desirable for the Air Force/NASA Joint NLS Program Office or its successor to have the necessary funds to assign high priority both to the new engine and to technologies and processes useful to existing vehicles.

Fourth, the NLS program or its successor, as well as other Air Force and NASA programs, is likely to emphasize improvements useful to many launcher types.[17] There will, however, be improvements unique to a specific existing launcher type. We recommend that some funds be made available to individual launch providers for improvements and even replacements of their own launchers and facilities. The funds could be used to lengthen or upgrade their own solid rocket motors and upper stages, improve engine performance, replace obsolete processing equipment, and even develop new engines or vehicles.[18] To have a multiplier effect, we also recommend that

---

[17]A possible exception is that the Air Force might emphasize improvements unique to Titan IV.

[18]A perhaps extreme example is the Delta Clipper under development at McDonnell Douglas. It is a single-stage-to-orbit vehicle with rocket engines alone. It aims to lift 25,000 lb into LEOs at $400 or less per pound of payload. David Webb, "Building a DC-3 for Space," *Space News*, January 6–12, 1992, p. 15. The point is not that McDonnell Douglas will likely succeed but that such company-funded launch projects should be encouraged.

the government require the recipients to match the grants with their own funds. The matching ratio remains to be specified.

Fifth, it is well recognized that the U.S. launch facilities are in dire need of repair and upgrade.[19] We agree with the findings of the Eastern Space and Mission Center (ESMC) that the United States "cannot afford to process, launch and track vehicles and payloads built with the technologies of the 21st century using instruments, facilities and methods developed 50 years ago."[20] Moreover, Air Force officials estimated that the repair and improvement of the Air Force launch facilities at Cape Canaveral could cost as much as $1 billion over the next decade.[21] Since these facilities are used by the civil and commercial sectors as well, the cost should not come solely from the Air Force budget.

DIRECT OR INDIRECT FEDERAL SUBSIDIES

The National Space Policy Directive of September 1990 stated that "U.S. Government Agencies will actively consider commercial space launch needs and factor them into their decisions on improvements in launch infrastructure and launch vehicles aimed at reducing cost, and increasing responsiveness and reliability of space launch vehicles." On the other hand, the U.S. National Space Policy released in November 1989 and the U.S. Commercial Space Policy Guidelines released in February 1991 both stated that "the United States will pursue its commercial space objectives without the use of direct Federal subsidies." Current U.S. space planners might not see potential conflicts between these two objectives, perhaps because they believe that a distinction between direct and indirect subsidies can be made. In fact, the issue of direct subsidy led the Bush Administration to reject a plan for providing money to states to im-

---

[19]See, for example, *ESMC 2005*, which describes what Eastern Space and Mission Center planners consider the launch facility requirements by the year 2005 and beyond. Also, the AIA Report to Samuel K. Skinner, *Commercial Space Launch Infrastructure: An Industry Perspective*, September 1990.

[20]Edward H. Kolcum, "Cape Canaveral Planners Seek Input on Sweeping Launch Changes," *Aviation Week & Space Technology*, March 18, 1991, p. 150.

[21]Vincent Kiernan, "AF Launch Pad Work Needed in 1990s," *Space News*, November 26–December 2, 1990, p. 1.

prove or build commercial space facilities.[22] We believe that the wording of "without the use of direct Federal subsidies" should be deemphasized, because the distinction is artificial. Subsidies are subsidies, regardless of whether they are direct or indirect. Foreign governments pay for the development of their launchers and their subsidies cannot be considered as merely indirect. The debates on direct or indirect subsidies can divert attention from the key issue— whether the proposed subsidies are beneficial to the United States. A justification for subsidies can be based on the simple fact that launch industries in all countries have long been subsidized by their governments. The United States should not be the sole exception. The United States should, however, be willing to work with other countries to reduce and eventually eliminate subsidies, especially if space launches are to be essentially commercialized.

In any case, there is one situation in which the U.S. government will always indirectly help the U.S. launch industry. That situation is when the U.S. government has unique launch requirements that pertain to national security. Quick launch responsiveness is one example. The launch industry would not make such an investment on its own for its commercial customers, because they do not need quick launches in a matter of days and, in any case, would not be willing to pay much for them. DoD needs to fund the infrastructure improvements for the quick launches it needs.

The processing time prior to on-pad assembly and integration can be shortened drastically by stocking and even pretesting satellites and launcher stages at the launch sites. The current bottleneck for quick launch is the long on-pad time. The planned on-pad times for Titan IV, Atlas II, and Delta II are about 110, 83, and 53 calendar days, respectively. These times are too unresponsive for most contingencies. Since Delta already has the shortest on-pad time among the three launch types and is equipped with the most substantial off-pad facilities to off-load further on-pad processing, it has the greatest potential to become a responsive vehicle. Future on-pad times as short as 14 days seem quite feasible.[23] The on-pad times for Atlas and

---

[22]Andrew Lawler, "Plan to Fund Commercial Launch Site Work Stalls in Senate," *Space News*, August 19–25, 1991, p. 7.

[23]Robert S. Fine, "Delta and ALS: Toward Efficient Launch Operations," McDonnell Douglas Space Systems Company, Cape Canaveral Air Force Station, Florida

Titan can also be reduced considerably.[24] The techniques used by the three launch providers to shorten the on-pad time will include performing more tests off-pad, streamlining on-pad test procedures, and running tasks in parallel. If DoD wants a rapid launch capability for medium, as opposed to small, satellites for emergencies, it has to provide the funding, since commercial customers do not have a strong need for such quick launches. This is not to say that responsiveness improvements would not benefit the commercial launch industry. They would.

(presented at the Conference of American Institute of Aeronautics and Astronautics on March 21, 1991).

[24]See, for example, R. J. Moberly, "Spaceport Florida Infrastructure Improvement Study," mid-term briefing, February 14, 1991, General Dynamics Space Systems Division. The study set a goal of reducing the Atlas/Centaur launcher on-pad time by about a third.

FINDINGS AND RECOMMENDATIONS

Our findings and recommendations are grouped into DoD space launch procurement and DoD steps to strengthen U.S. launch competitiveness.

DoD SPACE LAUNCH PROCUREMENT

Procurement Classification and Reliability Record

We classified launch procurement contracts into three types: government (GL), commercial-like (CLL), and commercial (CL). There is a view that GLs are preferable, because CLs or even CLLs are not as reliable. We, however, found that the launch data do not show, with high confidence, that different launch procurement types necessarily have different reliabilities.

Encouraging Commercialization by Reducing a Manager's Worries

Many DoD satellite and launch managers do not wish to reduce government oversight by using CLs, thus slowing the pace of launch commercialization. We propose that, whenever the DoD makes a comparative analysis to select a procurement type, the CL procurement option include a satellite/launcher backup at the launch site to reduce delay from a launch failure. If a CL procurement, including the backup cost, is still cheaper than a GL or CLL without a backup, CL should be considered. If a backup were found unnecessary, a CL procurement would be even cheaper. Also, we recommend that DoD consider systematically the inclusion of launch insurance on launchers and satellites in the launch contract.

Deletion of Undesirable Contract Features

The use of two prices, target and ceiling, in MLV-1 and -2 has forced the government to monitor costs and the contractor to provide cost data and to explain cost variances. Had a contract with a single fixed price been used, cost monitoring would have been unnecessary, because the government would pay the same price regardless of actual cost. Another cost-saving measure would be progress payments based on the passage of time instead of the portion of work accomplished. Finally, the typical price certification clause, which is meant to let the government benefit from future lower prices charged to other customers, might preclude a contractor from meeting the competition by lowering the price.

A Model to Help Decide Whether to Go Commercial

We developed a model for selecting a launch procurement type. Applying the model to existing launch programs, we found that our results are consistent with the Air Force's in not using CLL or CL to procure Titan IVs. The Air Force used a CLL procurement for MLV-1. In retrospect, our model results suggest that the Air Force could have saved money by using a CL. On the other hand, the lack of a CL record in 1987 made the choice of a CLL procurement for MLV-1 reasonable at that time. Our analytical results are consistent with the Air Force's CLL procurement for MLV-2 and the Navy's CL procurement for UHF Follow-ons.

Factors for Deciding Whether to Go Commercial

DoD can use the following set of questions to evaluate whether a particular DoD launch program should be procured commercially (CL instead of CLL and GL):

- Does the procurement selection model indicate a favorable break-even point for commercial launches?

- Do launch vehicles need to be modified?

- How seriously will launch delay affect the timeliness and quality of mission performance?

- If one decided to use commercial launches, would it be feasible and inexpensive to switch to CLLs or GLs if necessary? If so, what is the penalty at any given switch point?

Recommending That Commercial Procurement Be Included as an Option in MLV-3 RFP

The Air Force has incorporated many commercial features into MLV-1 and -2 and is planning to procure the important, upcoming MLV-3 for launching 20 GPS Follow-on satellites in a similar way—CLL. The Air Force could, however, better encourage commercialization by procuring MLV-3 commercially (CL) while incurring only a limited risk. Short of recommending CLs for MLV-3 outright, we suggest that the MLV-3 RFP contain CL and CLL options as well as various contractor liability options, because the Air Force will need pricing data to ascertain whether CL can have cost savings and whether the financial impact of launch failures can be limited.

Pace of Launch Commercialization

We consider an evolutionary approach to space launch commercialization to be both feasible and desirable. Regardless of the pace, there should always be room for a few launches to be procured differently. There is already a consensus that small launchers can be procured commercially. We now recommend that commercial procurement be considered for MLV-3. Titan IVs, on the other hand, are not yet ready for commercial procurement. Whether they should be commercially procured in the future depends on the commercial reliability record of Titan IIIs and on how well the commercialization of medium-lift launchers, such as Deltas and Atlases, fares.

DoD STEPS TO STRENGTHEN U.S. LAUNCH COMPETITIVENESS

A Justification for Subsidies

The statement that "the United States will pursue its commercial space objectives without the use of direct Federal subsidies," both in the November 1989 National Space Policy and the February 1991 Commercial Space Policy Guidelines, should be deemphasized in future directives. The distinction between direct and indirect subsidies is artificial. The debates on direct or indirect subsidies can divert the attention from the key issue—whether the proposed subsidies are beneficial to the United States A justification for subsidies can be based on the fact that launch industries in all countries have long been subsidized by their governments. The United States should not be the sole exception. The United States should, however, be willing to work with other countries to reduce and eventually eliminate launch subsidies.

Changes in DoD Launch Demand

Without the heavy SDI platforms anticipated in the past, the DoD's demand for VHLLVs (above 50,000 lb of payload to LEDs) could be low, whereas NASA's would be much higher. This key difference requires significant compromise in optimization of the engine and other design considerations for a joint Air Force/NASA launch program.

Launch Developmental Program Must Cover Medium-Lift Vehicles (10,000–30,000 lb to LEOs)

We consider that the most commercially relevant (MCR) range is the capability to lift 10,000 to 50,000 lb of payload into LEOs or 2000 to 10,000 lb into GSOs. Our greatest concern is that any new NLS-type family of vehicles might have a lower capacity bound at 30,000 lb for LEOs or 6000 lb to GSOs and thus miss a significant portion of the MCR range. Without government financial and other supports to improve and eventually replace vehicles in the MCR range, the U.S.

commercial launch industry will suffer greatly and possibly disappear, as competition is intensified by foreign, low-price launch providers and heavily subsidized newer launchers.

Government Support Needed to Compete in Commercial Heavy-Lift Vehicles (30,000–50,000 lb to LEOs)

Although Titan III can launch two geosynchronous communications satellites at a time, Martin Marietta no longer matches payloads to the same launcher for two different customers. This policy, in essence, positions Titan III launchers for a much smaller market segment of unusually large payloads. Ariane 5, beginning service by the mid-1990s, is expected to include dual-payload launches and to have a launch cost per pound that is 45 percent lower than the already highly competitive Ariane 4. With government support for cost reduction and performance improvement in existing vehicles, the U.S. launch industry believes it can be competitive with Ariane 5 until 2005, perhaps to 2010. We believe that the long-term solution is the early development of new vehicles in this 30,000-50,000-lb lift class.

Foreign Partners for Joint Development in Very-Heavy-Lift Launch Vehicles (Above 50,000 lb to LEOs)

This lift class is particularly suitable for an international joint development. Cost sharing in VHLLV development might leave adequate funds for the development of a new family of launch vehicles in the important MCR lift range (10,000 to 50,000 lb). The United States should actively seek foreign participation in a VHLLV venture.

Matching Funds for Improvements in Existing Vehicles

The NLS program or any successor, as well as other Air Force and NASA programs, is likely to emphasize improvements useful to many launcher types. There will, however, be improvements unique to a specific existing launcher type. We recommend that matching funds be made available to individual launch providers for improvements of their own launchers and facilities. The matching ratio remains to be specified. We also agree with assessments that U.S. launch facilities are in dire need of repair and upgrade. Otherwise, obsolete equipment will eventually degrade launch reliability, which is a key determinant in a customer's decision in selecting a launch provider.